MIX
Papier aus verantwortungsvollen Quellen
Paper from responsible sources
FSC® C105338

Sven Bodo Wirsing

Maximal nilpotent subalgebras II

A correspondence theorem within solvable associative algebras

With 242 exercises

Anchor Academic
Publishing

Wirsing, Sven Bodo: Maximal nilpotent subalgebras II: A correspondence theorem within solvable associative algebras. With 242 exercises, Hamburg, Anchor Academic Publishing 2017

Buch-ISBN: 978-3-96067-196-1
PDF-eBook-ISBN: 978-3-96067-696-6
Druck/Herstellung: Anchor Academic Publishing, Hamburg, 2017
Covermotiv: designed by freepik.com

Bibliografische Information der Deutschen Nationalbibliothek:
Die Deutsche Nationalbibliothek verzeichnet diese Publikation in der Deutschen Nationalbibliografie; detaillierte bibliografische Daten sind im Internet über http://dnb.d-nb.de abrufbar.

Bibliographical Information of the German National Library:
The German National Library lists this publication in the German National Bibliography. Detailed bibliographic data can be found at: http://dnb.d-nb.de

All rights reserved. This publication may not be reproduced, stored in a retrieval system or transmitted, in any form or by any means, electronic, mechanical, photocopying, recording or otherwise, without the prior permission of the publishers.

Das Werk einschließlich aller seiner Teile ist urheberrechtlich geschützt. Jede Verwertung außerhalb der Grenzen des Urheberrechtsgesetzes ist ohne Zustimmung des Verlages unzulässig und strafbar. Dies gilt insbesondere für Vervielfältigungen, Übersetzungen, Mikroverfilmungen und die Einspeicherung und Bearbeitung in elektronischen Systemen.

Die Wiedergabe von Gebrauchsnamen, Handelsnamen, Warenbezeichnungen usw. in diesem Werk berechtigt auch ohne besondere Kennzeichnung nicht zu der Annahme, dass solche Namen im Sinne der Warenzeichen- und Markenschutz-Gesetzgebung als frei zu betrachten wären und daher von jedermann benutzt werden dürften.

Die Informationen in diesem Werk wurden mit Sorgfalt erarbeitet. Dennoch können Fehler nicht vollständig ausgeschlossen werden und die Diplomica Verlag GmbH, die Autoren oder Übersetzer übernehmen keine juristische Verantwortung oder irgendeine Haftung für evtl. verbliebene fehlerhafte Angaben und deren Folgen.

Alle Rechte vorbehalten

© Anchor Academic Publishing, Imprint der Diplomica Verlag GmbH
Hermannstal 119k, 22119 Hamburg
http://www.diplomica-verlag.de, Hamburg 2017
Printed in Germany

Thank you Dad

Thank you Dad,
For everything you've done,
Thank you Dad,
From your thankful son.

Thank you Dad,
For just being there,
Thank you Dad,
For showing that you care.

Thank you Dad,
For showing us the way,
Thank you Dad,
For working hard every day.

Thank you Dad,
You're the very best,
Thank you Dad,
Because you're none like the rest.

Contents

Introduction		**7**
1	**Standard examples, symbols and notations**	**13**
2	**Radical algebras and the theorem of Xiankun Du**	**21**
	2.1 Radical algebras and central chains	21
	2.2 Results of Stephen Arthur Jennings, Hartmut Laue and Xiankun Du	23
	2.3 Standard examples	26
	2.3.1 The algebras of upper and lower triangular matrices	26
	2.3.2 Solomon algebras in characteristic zero	27
	2.3.3 Solomon-Tits algebras	28
	2.4 Open-ended questions and exercises	31
3	**Solvability**	**37**
	3.1 Solvability of the associated Lie algebra	37
	3.2 Solvability of the group of units	41
	3.3 The theorem of Sophus Lie and Borel subalgebras	43
	3.4 Standard examples	45
	3.4.1 Group algebras	45
	3.4.2 The algebras of upper and lower triangular matrices	46
	3.4.3 Solomon algebras in characteristic zero	46
	3.4.4 Solomon-Tits algebras	46
	3.5 Open-ended questions and exercises	53
4	**Carter subgroups of the group of units of solvable associative algebras**	**59**
	4.1 The pendant within the associated Lie algebra: Cartan subalgebras	59
	4.2 The determination of the Carter subgroups by Thorsten Bauer	60
	4.3 Further connections in the case of finite fields	65
	4.4 Standard examples	67
	4.4.1 Group algebras	67
	4.4.2 The algebras of upper and lower triangular matrices	70

      4.4.3   Solomon algebras in characteristic zero 71
      4.4.4   Solomon-Tits algebras 71
  4.5 Open-ended questions and exercises 74

5 The Fitting subgroup of the group of units of a solvable associative algebra   77

  5.1 The pendant within the associated Lie algebra: the nilradical   77
  5.2 The partner within the group of units: the Fitting subgroup . 77
  5.3 Standard examples 79
      5.3.1   The algebras of upper and lower triangular matrices . 79
      5.3.2   Solomon algebras in characteristic zero 79
      5.3.3   Solomon-Tits algebras 80
      5.3.4   Group algebras 80
  5.4 Open-ended questions and exercises 83

6 Maximal nilpotency in Lie algebras associated to solvable associative algebras   85

  6.1 Associativity 86
  6.2 Manifold centralizers 89
  6.3 Futile algebras 103
      6.3.1   Futility and radical complements 103
      6.3.2   Futility and unital subalgebras 110
  6.4 Finiteness of the number of isomorphism classes 114
  6.5 Cardinalities 117
  6.6 Examples 120
  6.7 Summary 122
  6.8 Open-ended questions and exercises 124

7 A correspondence theorem between maximal nilpotent subgroups and Lie subalgebras   131

  7.1 The correspondence theorem 131
  7.2 Open-ended questions and exercises 135

8 Maximal nilpotency in unit groups of solvable associative algebras   137

  8.1 A direct decomposition 137
  8.2 Manifold centralizers 138
  8.3 Finiteness of the number of isomorphism classes 148
  8.4 Cardinalities 149
  8.5 Examples 153
  8.6 Summary 155
  8.7 Open-ended questions and exercises 157

9	**Fischer subgroups, nilpotent projectors and injectors**	**163**
	9.1 Fischer subgroups	163
	9.2 The pendant for Lie algebras: Fischer subalgebras	164
	9.3 Nilpotent projectors	165
	9.4 The pendant for Lie algebras: nilpotent Lie projectors	166
	9.5 Nilpotent injectors	167
	9.6 The pendant for Lie algebras: nilpotent Lie injectors	168
	9.7 Open-ended questions and exercises	170

10 Outlook on series III   **173**

List of tables   **175**

List of figures   **177**

Bibliography   **177**

Index   **184**

Introduction

Within series I we have focussed on the following two main topics: the determination of the Cartan subalgebras and of the nilradical of the associated Lie algebra A° based on a finite-dimensional associative unitary algebra A. Both Lie substructures are maximal nilpotent in A°: Cartan subalgebras with respect to the subalgebra lattice and the nilradical with respect to the ideal lattice of A°. If the factor algebra by the nilradical of A is separable, then – by using the theorem of Wedderburn-Malcev – a radical complement T of $rad(A)$ in A exists. Based on this radical complement we were able to determine within series I the Cartan subalgebras and the nilradical of A° for several classes of algebras A. In particular, if A is solvable (which is the case of $A/rad(A)$ and T being commutative) we have proven that the centralizers of the radical complements – denoted by $C_A(T)$ – are exactly the Cartan subalgebras of A°. This results was proven originally by Thorsten Bauer within his dissertation [4]. In particular, all Cartan subalgebras of A° are associative subalgebras of A. The theorem of Wedderburn-Malcev is used to prove further that all Cartan subalgebras of A° are conjugated under the group $1 + rad(A)$. If we focus on the central part of T in A – which is $Z(A) \cap T$ – we have derived in series I additionally that this part is separable and that the nilradical of A° is the inner direct sum of $rad(A)$ and $Z(A) \cap T$. Cartan subalgebras are maximal Lie nilpotent subalgebras. If A is solvable, then the nilradical of A° is a maximal Lie nilpotent subalgebra, too.

Within this series we will enhance this theory of maximal nilpotent subalgebras of A° in the **solvable case** of A further. The following questions are the guidelines of this series related to the associated Lie algebra A° and also to the group of units $E(A)$ of A:

· In what way can we determine **all** maximal nilpotent Lie subalgebras of A°?

· Does a special or extremal position of the nilradical and the Cartan subalgebras exist among all maximal nilpotent Lie subalgebras of A°?

· In what way can we determine the Carter subgroups and the Fitting subgroup of $E(A)$? Is the Fitting subgroup a maximal nilpotent subgroup?

- In what way can we determine **all** maximal nilpotent subgroups of $E(A)$?

- Does a special or extremal position of the Fitting subgroup and the Carter subgroups exist among all maximal nilpotent subgroups of $E(A)$?

- Do structural connections exist between maximal nilpotent subalgebras of A° and maximal nilpotent subgroups of $E(A)$?

The intention of chapter 1 is to summarize special structures like group algebra, the Solomon algebra or the Solomon Tits algebra. These algebras are used to visualize the results within this work and to guide the reader within the exercises to a deeper insight of the proven results.

For the analysis of structural connections between maximal nilpotent subgroups and Lie subalgebras we will use the main result of chapter 2 frequently in this work: the theorem of Xiankun Du proven in 1992 based on radical algebras comprised that the upper central chain of the associated Lie algebra coincide with the upper central chain of the quasi regular group – or also called star or circle group (which is a generalization of the group of units) – in every step. In particular, the class of nilpotency of both structures is identical. This result was conjectured by Stephen Arthur Jennings 40 years ago and partly proven by Hartmut Laue in the eighties. Oftentimes, it is simpler to do calculations in the Lie algebra and not within the circle group. For example, radicals of associative algebras are radical algebras. In the context of maximal nilpotent substructures we use the result to derive a connection between the nilpotency classes of maximal nilpotent Lie subalgebras and maximal nilpotent subgroups. As an excursus at the end of chapter 2 we derive another application of the theorem of Xiankun Du. If we focus on the upper central chain of the circle group of a radical algebra and here on the factor groups of the $(n+1)$-th modulo the n-th center, then these factor groups are – by definition of the upper central chain – abelian groups. In the case of a radical algebra based on a field of positive characteristic p we can derive – by using the theorem of Xiankun Du – that these factor groups are indeed of exponent p. Applied to the group algebra – for which Adalbert Bovdi has published this result – the reader may prove this result within the exercises and experience the transfer of group theoretic questions to Lie theory.

As aforementioned, the guidelines of this work are connected to solvable associative algebras. The main focus will be to analyze structural properties of and connections between the associative and the associated Lie as well as the derived group structure in form of the group of units concerning maximal nilpotency. For the solvability itself a connection between these three structures is existing: we will prove within chapter 3 that the solvability for the associative algebra, its associated Lie algebra and its group of units based on a finite-dimensional associative unitary K-algebra (for a field K possessing at least 5 elements and $char(K) \neq 2$) are equivalent. This result was one incentive for our guidelines. As an excursus we focus at the end of chapter 3 on a connection between maximal solvable Lie subalgebras and maximal solvable subgroups: the so-called Borel subalgebras of A° – which are maximal solvable Lie subalgebras – are indeed associative unital subalgebras of A based for fields of characteristic zero. For proving this, we need a theorem of Sophus Lie and a result of Hartmut Laue concerning the associative algebra span. The group of units of the Borel subalgebras are solvable groups. Unfortunately, the proof that they are maximal solvable subgroups – which are so-called Borel subgroups – was not possible to perform. But we could prove that each Borel subalgebra leads to a new group of units. The reason is that the K-space generated by the group of units is the whole algebra. This approach – creating the group of units and the K-space generated by them – will often be useful within this work for describing and analyzing the connections between subalgebras and subgroups.

Thorsten Bauer has already analyzed one guideline of this work within his dissertation [4]: the determination of the Carter subgroups of the group of units of an unital finite-dimensional associative solvable algebra possessing a separable factor algebra by its nilradical. He has proven that the Carter subgroups – for a field possessing at least three elements – are exactly the group of units of the Cartan subalgebras of the associated Lie algebra. The assumption for the field is necessary to ensure that the algebra is generated by its group of units. Thus, the result of Thorsten Bauer can be reformulated as follows: the K-space generated by the Carter subgroups are exactly the Cartan subalgebras. Again, the concept of **creating the group of units and creating the K-space generated by the group of units** arise. Within the article [5] of Thorsten Bauer and Salvatore Siciliano concerning the determination of the Carter subgroups a result is proven which will be of significant importance later on in this work, too: the K-space generated of a nilpotent subgroup based on a finite-dimensional associative solvable K-algebra is Lie nilpotent.

The phenomenon of connecting Cartan subalgebras and Carter subgroups arise for the nilradical and the Fitting subgroup, too. We will prove within chapter 5 that both structures are connected via creating the group of units and the creating the K-space based on the group of units. The result of Thorsten Bauer and Salvatore Siciliano concerning the K-space generated by a nilpotent subgroup will be of significant importance for proving this connection.

The previous chapters have focussed on special and prominent examples of maximal nilpotent substructures within the group of units and the associated Lie algebra. In this chapter we analyze more generally the construction, determination and characterization of all maximal nilpotent Lie subalgebras. In a first step we prove – in analogy to the Borel subalgebras stated earlier (but based on a completely different argumentation) – that maximal Lie nilpotent subalgebras are unital associative subalgebras. Thus, we are able to use results of series I concerning these special associative subalgebras: the inner structure of these associative subalgebras M of A is presentable as the inner direct sum of its nilradical $rad(M)$ (which is contained in $rad(A)$ by using the solvability of A) and the unique and central radical complement $VSEP(M)$ consisting of fully separable elements: $M = rad(M) \oplus VSEP(M)$. The theorem of Wedderburn-Malcev lets us derive that $VSEP(M)$ is contained in a radical complement T of $rad(A)$ in A. Based on the inner structure of M and the radical complement T we can prove that a Lie nilpotent associative subalgebra M is maximal Lie nilpotent if and only if the centralizer conditions $C_{rad(A)}(VSEP(M)) = rad(M)$ and $C_T(rad(M)) = VSEP(M)$ are valid. A simple but remarkable consequence is that maximal nilpotent Lie subalgebras satisfy the double-centralizer conditions $C_{rad(A)}(C_T(rad(M))) = rad(M)$ and $C_T(C_{rad(A)}(VSEP(M))) = VSEP(M)$. For determining all maximal Lie nilpotent subalgebras we use these centralizer and double-centralizer properties: we start with an unital subalgebra C of T and calculate the double-centralizer $C_T(C_{rad(A)}(C))$. We proceed by calculating the double-centralizer of the double-centralizer again and again. This process must be stable of finite many steps because of the finite dimension of A. If the process is stable, then the resulting subalgebra $\hat{C}$ in T combined with the direct summand $C_{rad(A)}(\hat{C})$ is maximal Lie nilpotent. The dual process – beginning with a subalgebra of $rad(A)$ – leads also to maximal Lie nilpotent subalgebras, but not to new ones. A natural question is to determine the number of steps after which the double-centralizing is stable. The answer is simple: not from the beginning but after the first double-centralizing. Thus, we have to use the double-centralizing on the lattice of unital subalgebras of T resp. the lattice of subalgebras of $rad(A)$ once and construct as already described all maximal Lie nilpotent subalgebras. The nilradical and the Cartan subalgebras have an extremal position among all maximal Lie nilpotent subalgebras. The component of the nilradical resp. Cartan subalgebras in T is central in A resp. the whole radical complement. Within the nilradical its extremely large resp. small (and therefore dual). For all other maximal nilpotent subalgebras the part in T resp. $rad(A)$ is situated between these two values. By using another radical complement only isomorphic copies of maximal nilpotent Lie subalgebras arise (based on the theorem of Wedderburn-Malcev). Hence, all isomorphic classes of maximal nilpotent subalgebras can be bounded by the number of unital subalgebras of a fixed radical complement T. This number is finite because T is separable and commutative: T is a so-called futile algebra. We prove this statement within a separate section and estimate this number by the upper bound $B(dim_K(T))$ – which are the so-called Bell numbers.

In chapter 7 we present a bijective connection between maximal nilpotent Lie subalgebras and maximal nilpotent subgroups. It becomes apparent that – as already stated for the Cartan subalgebras and the Carter subgroups resp. the nilradical and the Fitting subgroup – there is a general connection between maximal nilpotent substructures: the group of units of maximal Lie nilpotent subalgebra (which is indeed an unital associative subalgebra) is a maximal nilpotent subgroup and the K-space generated by a maximal nilpotent subgroup is a maximal nilpotent Lie subalgebra (Here we will use the already mentioned result of T. Bauer and S. Siciliano again.). In addition, this connection is bijective: the functions $E(\cdot)$ – creating the group of units – and $\langle\cdot\rangle_K$ – creating the K-space generated by the group of units – are inverse to each other. By using the theorem of Xiankun Du we derive the more deeper insight that the classes of nilpotency of two connected maximal nilpotent substructures are identical. The results presented in chapter 6 can be transferred by using this connection to maximal nilpotent subgroups which is the content of chapter 8.

Thus, in analogy to chapter 6 we describe within chapter 8:

· the inner structure of the maximal nilpotent subgroups as direct products of unipotent and central, fully separable elements,

· the characterization of maximal nilpotent subgroups by manifold centralizers,

· the determination of all maximal nilpotent subgroups by double-centralizing all subgroups of $E(T)$ and combining the centralized unipotent part to it,

· the dual determination of all maximal nilpotent subgroups by double-centralizing all subgroups of $1 + rad(A)$ and combining the centralized fully-separable part to it,

· the extremal position of the Carter subgroups and the Fitting subgroup among all maximal nilpotent subgroups of $E(A)$,

· the behavior of the maximal nilpotent subgroups by changing the radical complement and

· the finiteness of the number of isomorphic classes of maximal nilpotent subgroups which can be bonded by Bell numbers.

The last chapter is dedicated to other prominent maximal nilpotent subgroups which are the so-called nilpotent injectors, nilpotent projectors and the Fischer subgroups. We will prove that they coincide with the Carter subgroups resp. the Fitting subgroup. Afterwards these prominent maximal nilpotent substructures are also defined for Lie algebras (nilpotent Lie injectors, nilpotent Lie projectors and Fischer subalgebras), and we will prove that they coincide with the Cartan subalgebras and the Lie nilradical. Posthumous, we derive the result that the group of units of the Fischer subalgebras, the nilpotent Lie projectors and the nilpotent Lie injectors are exactly the Fischer subgroups, the nilpotent projectors and injectors. Vice versa, the K-space generated by them is exactly the Fischer subalgebras, the nilpotent Lie projectors and the nilpotent Lie injectors.

As stated earlier we illustrate our results by using standard examples. These are mainly group algebras, the algebras of upper and lower triangular matrices over a field, the Solomon algebra in characteristic zero and the Solomon-Tits algebra. Within the first chapters these examples are investigated on a high detailed level, but within the last four chapters we use them only exemplary. A detailed analysis needs a deeper insight, and the author decided not to disconnect the reader from the general theory too far, but to do this analysis in series III.

Some applications are also transferred to the exercises at the end of each section or chapter. There are some exercises included enhancing the theory presented so far such that the reader can experience a deeper insight. In addition, at the beginning of each exercise series some open-ended topics are included which can be used by the reader – and also by the author – to do additional researches within this theory. The author has included some manually created graphics – mostly so called Hasse diagrams – to visualize the main results of this work.

Excercise 1 *What are the answers for the guidelines of this work?*

Chapter 1

Standard examples, symbols and notations

This chapter has a preliminary function by summarizing those monoids, groups, associative and Lie algebras which will arise frequently in this work. They will be used as examples for the proven theorems as well as for the exercises in which the reader shall apply the general results to them. In addition, we list the symbols and notations used in this series.

Sets and numbers

Let A, B, T be sets and $i, n, k \in \mathbb{N}_0$. We use the following symbols linked to set and number theoretical topics:

- $\emptyset$ - the empty set
- $A \cap B$ - intersection of A, B
- $A \cup B$ - union of A, B
- $A \setminus B$ - difference of A, B
- $A \times B$ - cartesian product of A, B
- $P(A)$ - power set of A
- $\underline{n}$ - the first n natural numbers
- $\underline{n}_0$ - the first n natural numbers including zero
- $p(n)$ - the number of partitions of n
- $n!$ - factorial of n
- $|A|$ - number of elements of A

- $S(n,k)$ - the kth-Stirling number of n
- $B(n)$ - the nth-Bell number
- $\binom{n}{k}$ - n choose k
- $\binom{T}{i}$ - the set of subsets of order i of T
- $\equiv$ - equivalent
- mod - modulo.

Groups and monoids

Let $p \in \mathbb{P}$, $n \in \mathbb{N}$, N be a set, M a monoid, G a group, N a normal subgroup of G, $a, b \in G$, $U, V \leq G$, A an associative unitary K-algebra, $c \in K$ and q a prime power number. The following monoids, groups and symbols are used:

- $st(G)$ - solvable class of G
- $cl(G)$ - nilpotency class of G
- $(Z_n(G))_{n \in \mathbb{N}}$ - ascending central chain of G
- $(G^{(n)})_{n \in \mathbb{N}}$ - descending central chain of G
- $(\gamma_n(G))_{n \in \mathbb{N}}$, $(G^{[n]})_{n \in \mathbb{N}}$ - commutator or derived series of G
- $\star$, $\star_c$ - star or circle composition
- $[a, b]$ - commutator of a, b
- $[U, V]$ - commutator of U, V
- a^{-1} - inverse element of a
- a^b - conjugate element of a with b
- $1 + rad(A)$ - normalized units
- $C_U(V)$ - centralizer of V in U
- $N_U(V)$ - normalizer of V in U
- $\mathcal{G}$ - class of groups
- G' - derived subgroup of G
- $(F^n(G))_{n \in \mathbb{N}}$ - Fitting series of G

- G/N - factor group of G by N
- $\mathcal{SU}_{E(A)}$ - set of solvable subgroups of $E(A)$
- Π_n - ordered set partitions
- $\wedge_n$ - special composition on Π_n
- $exp(G)$ - exponent of G
- $U_\mathcal{U}$ - unipotent factor of U
- $U_\mathcal{V}$ - fully-separable factor of U
- $O_p(G)$ - p-core of G
- $\mathcal{G}(U), \mathcal{U}_{E(T)}$ - subgroups of $E(T)$ possessing the double-centralizer property
- U_T - order of $\mathcal{G}(U)$
- $\mathcal{G}_{DJ}$ - subgroups of $1 + rad(A)$ possessing the double-centralizer property
- $\mathcal{G}_{UT}$ - maximal nilpotent subgroups of $E(A)$ linked to $E(T)$
- $\mathcal{G}_U$ - maximal nilpotent subgroups $E(A)$
- $\mathcal{E}(A)_M$ - maximal nilpotent subgroups of $E(A)$
- a^- - quasiregular inverse element
- $x^{(a)}$ - quasiregular conjugate element
- $\mathbb{N}$ - natural numbers
- $\mathbb{N}_0$ - natural numbers containing zero
- D_{2n} - dihedral group of order $2n$
- Q_{4n} - quaternion group of order $4n$
- SD_{2^n} - semi-dihedral group of order 2^n
- S_n - symmetric group of degree n
- A_n - alternating group of degree n
- $GL(n,q)$ - general linear group of degree n over $GF(q)$
- C_n or Z_n - cyclic group of order n
- $E(A)$ - group of units of A

- $Q(A)$ - quasiregular group of A
- $\times$ - direct products of groups
- $\ltimes$ - semidirect product of groups.

General algebra constructions

Let A be an algebra, $S \subseteq A$, S finite, K a field, G a group, I an ideal, M a monoid, $n \in \mathbb{N}$, $\alpha \in S_n$ and $T \subseteq A$. The following general algebra constructions and symbols are used:

- $dim_K(A)$ - K-dimension of A
- $\mathcal{A}$ - class of associative algebras
- $\mathcal{A}_1$ - class of associative unitary algebras
- $st(A)$ - solvable class of A
- $cl(A)$ - class of nilpotency of A
- $D(\alpha)$ - defect class of α
- a_{nil} - nilpotent part of the generalized Jordan decomposition
- a_{vsep} - fully-separable part of the generalized Jordan decomposition
- $\overline{S}$ - sum of all elements of S
- A' - derivation of A
- A^n - nth associative power of A
- $C_A(T)$ - centralizer of T in A
- a_T - sum of a_i, $i \in T$, $T \subseteq \underline{n}$, $a_1, \cdots a_n \in A$
- $\otimes$ - tensor product of algebras
- $\times$ - direct products of algebras
- $\oplus$ - direct sum of algebras
- $\ltimes$ - semidirect product of algebras
- A/I - factor algebra of A by the ideal I
- KG - group algebra of the group G and the field K
- KM - monoid algebra of the monoid M and the field K

- $A^{n\times n}$ - algebra of $n \times n$-matrices over A
- A° - associated Lie algebra of A
- $\langle T \rangle_K$ - K-linear span of T
- $\langle T \rangle_\mathcal{A}$ - subalgebra generated by T
- $\langle T \rangle_{\mathcal{A}_1}$ - unital subalgebra generated by T
- A^K - adjunction of an unit to A
- A^{op} or A^- - opposite or inverse algebra of A
- $(A \times A; \odot)$ - zero extension of A
- $gl(n, K)$ - identical to $(K^{n \times n})^\circ$
- eAe - identical to $\{eae \mid a \in A\}$ for an idempotent e
- $Aug(KG)$ - augmentation ideal of KG.

Commutative associative algebras

Let $n \in \mathbb{N}$ and K be a field. The following commutative associative algebras and symbols are used:

- $D(n, K)$ the set of diagonal matrices in $K^{n \times n}$
- $e_1, \cdots, e_n$ - primitive orthogonal idempotents in $D(n, K)$
- e_T - sum of all e_i, $i \in T \subseteq \underline{n}$
- K^n - n-tuple space over the field K
- $K[a]$ - smallest subalgebra containing a and K
- $VSEP(\cdot)$ - set of fully-separable elements
- $char(K)$ - characteristic of the field K
- n_K - is identical to $\sum_{i=1}^{n} 1_K$
- $\mathbb{Z}$ - the set of integers
- $K[t]$ - polynomial algebra over K in one variable t
- $K[t_1, \ldots, t_n]$ - polynomial algebra over K in the variables $t_1, \ldots, t_n$.

Fields and skew fields

Let p be a prime number, $n \in \mathbb{N}$ and $(K; L)$ a field extension. The following fields, skew fields and symbols are used:

- $\mathbb{Q}$ - rational number field
- $\mathbb{R}$ - real number field
- $\mathbb{C}$ - complex number field
- $\mathbb{H}$ - real quaternion algebra
- $GF(p^n)$ - finite field possessing p^n elements
- $GF(q)$ - notation for $GF(p^n)$ and $q = p^n$
- $A(a, b)$ - generalized quaternion algebra
- $K(a)$ - smallest subfield in L containing a and K
- ω_d - primitive dth root of unity
- $K(t)$ - field of fractions over K in one variable t
- $K(t_1, \ldots, t_n)$ - field of fractions over K in the variables $t_1, \ldots, t_n$.

(Central-) simple associative algebras

Let K be a field, D a division algebra and $n \in \mathbb{N}$. The following (central-) simple associative algebras and symbols are used:

- e_{ij} - base matrices of $K^{n \times n}$
- det - determinant function
- tr - trace function
- $K^{n \times n}$ - $n \times n$-matrices over K
- $D^{n \times n}$ - $n \times n$-matrices over D
- $A(a, b)$ - generalized quaternion algebra.

Semisimple associative algebras

The following semisimple associative algebras are used:

- right artian associative algebras A for which $rad(A) = 0$ is valid
- $\times$ - direct products of simple algebras
- $A/rad(A)$ - the factor algebra by the nilradical of an associative right artian algebra.

Nilpotent associative algebras

Let A be an associative algebra, K a field, p a prime number, $n \in \mathbb{N}$ and G a group. The following nilpotent associative algebras are used:

- $rad(A)$ - nilradical of A
- $J(A)$ - Jacobson radical of A
- $s\delta_{u,n}$ - algebra of strict lower triangular matrices of $K^{n \times n}$
- $s\delta_{o,n}$ - algebra of strict upper triangular matrices of $K^{n \times n}$
- $Aug(KG)$ - augmentation ideal of KG based on a p-group G and $char(K) = p$.

Solvable associative algebras

Let $n \in \mathbb{N}$ and K be a field. The following solvable associative algebras are used:

- $K\Pi_n$ - Solomon-Tits algebra (see e.g. [77])
- D_n - Solomon algebra in the case $char(K) = 0$ (see e.g. [4])
- $\delta_{u,n}$ - algebra of lower triangular matrices of $K^{n \times n}$
- $\delta_{o,n}$ - algebra of upper triangular matrices of $K^{n \times n}$
- KG - group algebra based on: $char(K) = p$ and G possesses a normal p-Sylow subgroup and an abelian p'-Hall subgroup.

Lie algebras

Let $n \in \mathbb{N}$, K be a field, L a Lie algebra, $H \subseteq L$, $x \in L$ and A an associative algebra. The following Lie algebras and symbols are used:

- $\circ$ - associated Lie composition
- A° - associated Lie algebra
- $ad(x)$ - adjoint representation x
- $L_0(ad(H))$ - Fitting null component of L with resp. to $ad(H)$
- $\mathcal{B}_A$ - Borel subalgebras of A°
- $st(L)$ - solvable class of L

- $cl(L)$ - class of nilpotency of L
- $(Z_n(L))_{n\in\mathbb{N}}$ - ascending central chain of L
- $(L^{(n)})_{n\in\mathbb{N}}$ - descending central chain of L
- $(L^{[n]})_{n\in\mathbb{N}}$ - descending commutator or derived series of L
- $\mathcal{L}$ - class of Lie algebras
- $nil(L)$ - nilradical of L
- $\mathcal{M}(T)$, $\mathcal{A}_{DT}$ - subalgebras of T possessing the double-centralizer property
- m_T - order of $\mathcal{M}(T)$
- $\mathcal{A}_{DJ}$ - subalgebras of $rad(A)$ possessing the double-centralizer property
- $\mathcal{A}_{MT}$ - maximal nilpotent Lie subalgebras of A° linked to T
- $\mathcal{A}_M$ - maximal nilpotent Lie subalgebras of A°
- L' - derivation of L
- t_M, u_M - unit group and K-space creation
- $\mathfrak{J}_i(I)$ - special sequence of subalgebras and subgroups in $rad(A)$
- $\mathfrak{T}_i(C)$ - special sequence of subalgebras and subgroups in T.

Chapter 2

Radical algebras and the theorem of Xiankun Du

Within this chapter we focus on radical algebras. For these algebras a deep nilpotent connection between the associated Lie algebra and the circle or adjoint group proven by Xiankun Du is presented. This chapter is designed based on the manuscript of Hartmut Laue in [40]. Within this manuscript results of the diploma thesis of Karsten Scholz are used (see [58]). Proofs are available in chapter 4 in [40] and not stated here.

Based on Du's theorem the nilpotency classes of the associated Lie algebra and the adjoint circle group resp. the group of units are identical. Thus, the determination of the nilpotency class of the circle group can be handled by calculations purely within the associated Lie algebra and vice versa. In some applications it is much more easier to do the calculation within the Lie algebra as within the circle group. For this transfer principle some applications are presented within the exercises. In addition, one theorem about the p-power structure of the circle group in characteristics p is proven by using this transfer principle to the Lie algebra.

Radical algebras and their analysis concerning nilpotency of the associated Lie algebra and the circle group will play an important role later on in this work: the correspondence theorem between maximal nilpotent Lie subalgebras and subgroups.

2.1 Radical algebras and central chains

Definition and remark 1 Let A be an associative K-algebra. By $rad(A)$ resp. $J(A)$ we symbolize the nilradical resp. the Jacobson radical of A. If A is right or left artian, then Gottfried Köthe has proven that both radicals coincide and are nilpotent.[1] The associative nilpotency class is symbolized by $cl(A)$. A is called radical algebra, if $A = J(A)$ is valid. A is a nil algebra, if $A = rad(A)$ is true. $rad(A)$ and $J(A)$ are radical algebras.

The Lie algebra associated to A is symbolized by A° equipped with the multiplication $a \circ b := ab - ba$ for all $a, b \in A$. The upper central chain of A° is defined recursively by $Z_0(A^\circ) := \{0\}$ and $Z_n(A^\circ) := \{z \mid z \in A, \forall a \in A : z \circ a \in Z_{n-1}(A^\circ)\}$ for all $n \in \mathbb{N}$. A is Lie nilpotent, if a natural number $n \in \mathbb{N}$ exists such that $A = Z_n(A^\circ)$ is valid. The minimal n possessing this property is called the class of nilpotency of A° – symbolized by $cl(A^\circ)$. The lower central chain is defined recursively by $(A^\circ)^{(0)} := A$ and $(A^\circ)^{(n)} := (A^\circ)^{(n-1)} \circ A$ for all $n \in \mathbb{N}$. A is Lie nilpotent if and only if the lower central chains reaches the null space after finite many steps. The minimal number of these steps is again the class of nilpotency. Sufficient – and often used within applications – for the Lie nilpotency is the associative nilpotency, e.g. for $rad(A)$ if A is right artian. In [40] it is proven by Hartmut Laue that all members of the upper Lie central chain are associative subalgebras.

For all $a, b \in A$ we define (as original defined by Bartel Leendert van der Waerden) $a \star b := a + b + ab$, and we call $\star$ the circle or star composition on A. A is a monoid based on the composition $\star$, and 0 is its unit element. The group of units based on this monoid is called the star group or quasi regular group of A – symbolized by $Q(A)$. The elements of $Q(A)$ are

---

[1] Gottfried Maria Hugo Köthe (born 25 December 1905 in Graz; died 30 April 1989 in Frankfurt) was an Austrian mathematician working in abstract algebra and functional analysis. In 1923 Köthe enrolled in the University of Graz. He started studying chemistry, but switched to mathematics a year later after meeting the philosopher Alfred Kastil. In 1927 he submitted his thesis Contributions to Finslers foundations of set theory and was awarded a doctorate. After spending a year in Zürich working with Paul Finsler, Köthe received a fellowship to visit the University of Göttingen, where he attended the lectures of Emmy Noether and Bartel van der Waerden on the emerging subject of abstract algebra. He began working in ring theory and in 1930 published the Köthe conjecture stating that a sum of two left nil ideals in an arbitrary ring is a nil ideal. By a recommendation of Emmy Noether, he was appointed an assistant of Otto Toeplitz in Bonn University in 1929 to 1930. During this time he began transition to functional analysis. He continued scientific collaboration with Toeplitz for several years afterward. Köthes Habilitationsschrift Skew fields of infinite rank over the center, was accepted in 1931. He became Privatdozent at University of Münster under Heinrich Behnke. During World War II he was involved in coding work. In 1946 he was appointed the director of the Mathematics Institute at the University of Mainz and he served as a dean (1948 to 1950) and a rector of the university (1954 to 1956). In 1957 he became the founding director of the Institute for Applied Mathematics at the University of Heidelberg and served as a rector of the university (1960 to 1961). Köthes best known work has been in the theory of topological vector spaces. In 1960, volume 1 of his seminal monograph topological vector spaces was published (the second edition was translated into English in 1969). It was not until 1979 that volume 2 appeared, this time written in English. He also made contributions to the theory of lattices.

called quasi regular, the inverse of $a \in Q(A)$ will be denoted by a^-, and the conjugated to a by b is symbolized by $a^{(b)} := b^- \star a \star b$. Every nilpotent element is quasi regular, and thus for every nil associative algebra A the identity $A = Q(A) = rad(A)$ is valid. If $Q(A) = A$ is true, then we use the symbol $A^\star$ for $Q(A)$. $rad(A)$ is a group based on $\star$. The Jacobson radical is a group based on $\star$, too. The upper central chain of $Q(A)$ is recursively defined by $Y_0(Q(A)) := \{0\}$ and $Y_n(Q(A)) := \{y \mid y \in Q(A), \forall a \in Q(A) : [y,a] \in Y_{n-1}(Q(A))\}$ for all $n \in \mathbb{N}$. The commutator $[y,a]$ is defined by $y^- \star y^{(a)}$ for all $y, a \in Q(A)$. $Q(A)$ is nilpotent, if the upper central chain of $Q(A)$ reaches $Q(A)$ in finite many steps. The minimal number of these steps is called the class of nilpotency of $Q(A)$ – symbolized by $cl(Q(A))$. The lower central chain of $Q(A)$ is defined recursively by $Q(A)^{(0)} = Q(A)$ and $Q(A)^{(n)} := [Q(A)^{(n-1)}, Q(A)]$ for all $n \in \mathbb{N}$. $Q(A)$ is nilpotent if and only if the lower central chain reaches the trivial subgroup after finite many steps. The minimal number of these steps is again the class of nilpotency. Within the literature $\gamma_k(G)$ is used for the k-th member of the lower central chain of a group G.

The group $Q(A)$ acts per conjugation on the additive group of A. For this operation another central chain can be defined recursively by $X_0(Q(A)) := \{0\}$ and $X_n(Q(A)) := \{x \mid x \in Q(A), \forall a \in Q(A) : x^{(a)} - x \in X_{n-1}(Q(A))\}$ for all $n \in \mathbb{N}$.

In addition, we define for a radical algebra A inductively $W_0(A) := \{0\}$ and $W_n(A) := \{w \mid w \in A, \forall a \in A : a^{(w)} - a \in W_{n-1}(A)\}$ for all $n \in \mathbb{N}$.

Now we demonstrate how the nilpotency and the upper central chains of $A^\star$ and A° are connected for radical algebras. ◇

2.2 Results of Stephen Arthur Jennings, Hartmut Laue and Xiankun Du

A first insight about a nilpotent connection between the group $A^\star$ the Lie algebra A° was proven by Stephen Arthur Jennings in 1955 (see [30]) for radical algebras:

Theorem 1 *(Jennings, 1955) Let A be a radical algebra. $A^\star$ is nilpotent if and only if A° is nilpotent.*◇

Stephen Arthur Jennings[2] conjectured for radical algebras A that the nilpotency classes of $A^\star$ and A° coincide. In 1984 Hartmut Laue has proven some

---
[2]Stephen Arthur Jennings (May 11, 1915 to February 2, 1979) was a mathematician who made significant breakthroughs in the study of modular representation theory (1941). His advisor was Richard Brauer, and his student Rimhak Ree discovered two infinite

aspects of this conjecture in [41]:

Theorem 2 *(Laue, 1984) Let A be an associative K-algebra. The following statements are valid:*

(i) If $A = Q(A)$ and A is Lie-nilpotent, then $A^\star$ is nilpotent and $cl(A^\star) \leq cl(A^\circ)$ is valid.

(ii) Let A be nil, K a field and A° nilpotent. $Z_n(A^\circ) = Y_n(A^\star)$ is valid for all $n \in \mathbb{N}_0$.

(iii) If $Q(A) = A$ is valid, then $Z_2(A^\circ) = Y_2(A^\star)$ is true.⋄

Hartmut Laue conjectured that for a radical algebra A the ascending central chains of $A^\star$ and A° are identical in every step. Xiankun Du has proven this conjecture in 1992 (see [14]). Thus, the conjecture of Stephen Arthur Jennings was proven, too.

Theorem 3 *(Du, 1992) Let A be a radical algebra. For all $n \in \mathbb{N}_0$ the identity $Z_n(A^\circ) = Y_n(A^\star)$ is valid.*

In particular, the conjecture of Jennings is true: $A^\star$ is nilpotent if and only if A° is nilpotent. If $A^\star$ or A° is nilpotent, then $cl(A^\circ) = cl(A^\star)$ is valid.⋄

Further analysis performed by Karsten Scholz and Hartmut Laue (see [40] and [58]) yields to the following main theorem:

Main theorem 1 *(Laue, Scholz, 1996) Let A be a radical algebra. For all $n \in \mathbb{N}_0$ the identity $Z_n(A^\circ) = Y_n(A^\star) = X_n(A) = W_n(A)$ is valid.*

In particular, every member of these four central chains are additive closed, associative subalgebras of A, Lie ideals of A° and normal subgroups of $A^\star$, and they are invariant under all automorphism of $A^\star$ and A° and under all $A^\star$-module automorphism of the additive group of A.⋄

---

series of finite simple groups known as the Ree groups. He was an editor of Mathematics Magazine and an acting president of the University of Victoria. Stephen was born in Walthamstow, England and immigrated to Canada with his family in 1928. He had been receiving scholarships in England and these were transferred to Canada. He finished his high school education in Toronto and in September 1932 he went to University College in Toronto. In 1939 he received his PHD from the University of Toronto. He married Dorothy Freeda Rintoul (University of Western University - B.A. and University of Toronto M.A. Psychology) in 1939. On November 14, 1942 he became a member of the Zeta Psi fraternity. When he was made a professor, he was the youngest professor ever appointed in Canada. In 1944 Stephen was appointed 2nd Lieutenant (Paymaster) in Canadian Army. While in Vancouver, teaching at the University of British Columbia, Stephen and Dorothy established their family, two children, Judith Anne Jennings and James Stephen Jennings. Stephen was Dean of Graduate Studies at the University of Victoria and the Head of the Math Department there.

We close this section by proving an application based on the theorem of Xiankun Du concerning the factor groups along the ascending central chain of the circle group of a radical algebra in positive characteristic p: they are elementary-p-abelian. The proof demonstrates the transfer of group-theoretic topics to questions within Lie algebras.

Let T be a set and $i \in \mathbb{N}_0$. By $\binom{T}{i}$ we denote the set of all subsets of T possessing exactly i elements. With respect to the natural order on $\lfloor T \rfloor$ and T exactly one monotone bijection from $\lfloor T \rfloor$ onto T exists, and we symbolize this map by φ_T.

The next proposition is straightforward to be proven by an induction argument and therefor maybe handled by the reader as an exercise:

Proposition 1 *Let A be an associative K-algebra, $n \in \mathbb{N}$ and $x_1, \ldots, x_n \in A$. The statement $x_1 * \cdots * x_n = \sum_{i=1}^{n} \sum_{T \in \binom{\underline{n}}{i}} x_{(1\varphi_T)} \cdots x_{(i\varphi_T)}$ is valid.*⋄

Corollary 1 *Let A be an associative K-algebra, $n \in \mathbb{N}$ and $a \in A$. The following statements are valid:*

(i) $\underbrace{a * \cdots * a}_{n-fold} = \sum_{i=1}^{n} \binom{n}{i}_K a^i$

(ii) *If p is a prime number and $char(K) = p$ is valid, then* $\underbrace{a * \cdots * a}_{p^n-fold} = a^{(p^n)}$ *is true.*

Proof. ad(i): By using proposition 1 we derive
$$\underbrace{a * \cdots * a}_{n-fold} = \sum_{i=1}^{n} \sum_{T \in \binom{\underline{n}}{i}} a^i = \sum_{i=1}^{n} |\binom{\underline{n}}{i}|_K a^i = \sum_{i=1}^{n} \binom{n}{i}_K a^i.$$
ad(ii): For all $i \in \underline{p^n - 1}$ it is well-known (see e.g. [18])that the prime number $p = char(K)$ is a divisor of $\binom{p^n}{i}$. Thus part (ii) is a consequence of part (i).⋄

The proof of the following proposition is straightforward to be executed and thus maybe done as an exercise by the reader:

Proposition 2 *If A is an associative K-algebra and $x, y \in A$, the following statements are valid:*

(i) $\forall n \in \mathbb{N} : x \circ \underbrace{y \circ \cdots \circ y}_{n-fold} = \sum_{k=0}^{n} \binom{n}{k}_K (-1_K)^k y^k x y^{n-k}$.

(ii) If p is a prime number and $char(K) = p$ is valid, then the following identity is true: $x \circ \underbrace{y \circ \cdots \circ y}_{p-fold} = x \circ y^p.\diamond$

Theorem 4 *Let p be a prime number, K a field, $char(K) = p$ and A a K-radical algebra. For all $n \in \mathbb{N}$ the identity $exp(Z_{n+1}(A^*)/Z_n(A^*)) = p$ is valid.*

Proof. By using the theorem of Xiankun Du (see theorem 3) for all $n \in \mathbb{N}$ the sets $Z_n(A^*)$ and $Z_n(A^\circ)$ are identical. Based on corollary 1 for all $a \in A$ the identity $\underbrace{a * \cdots * a}_{p-fold} = a^p$ is valid. The theorem is proven by using proposition 2.$\diamond$

We remark that the exponent of the center is not p in general. The structure of the center is analyzed by the author within [76] for a p-group and a (finite) field of characteristic p. Within this context theorem 4 is applicable, and, in addition, the theorem of Xiankun Du can be used to calculate the class of Lie nilpotency for the nilradical. Some examples are included within the exercises in which the reader can experience the connection between group and Lie theory.

In the next chapter we apply the latter results to our standard examples.

2.3 Standard examples

2.3.1 The algebras of upper and lower triangular matrices

Let K be a field and $n \in \mathbb{N}$. The algebras of upper and lower triangular matrices are examples of solvable associative K-algebras possessing a factor algebra by its nilradical which is isomorphic to a n-fold product of the base field. In addition, it can be proven that every radical complement is self-centralizing.

The algebra of lower triangular matrices of $K^{n \times n}$ - symbolized by $\delta_{u,n}$ - possesses as nilradical the subalgebra of strict lower triangular matrices - symbolized by $s\delta_{u,n}$. The nilradical is of dimension $\sum_{i=1}^{n-1} i = \frac{1}{2}(n-1)n$. The set of diagonal matrices $D(n, K)$ is a radical complement of dimension n, and it is self-centralizing.

The algebra of upper triangular matrices of $K^{n \times n}$ - symbolized by $\delta_{u,n}$ - possesses as nilradical the subalgebra of strict upper triangular matrices - symbolized by $s\delta_{u,n}$. The nilradical is of dimension $\sum_{i=1}^{n-1} i = \frac{1}{2}(n-1)n$. Again, the set of diagonal matrices $D(n, K)$ is a radical complement.

It is well-known (see e.g. [38]) that for the associative nilpotency classes the following identities are valid: $cl(s\delta_{o,n}) = cl(s\delta_{u,n}) = n$.

Within [77] it is proven for an associative algebra possessing a self-centralizing radical complement and a radical factor algebra isomorphic to a n-fold direct product of the base field that the associative powers of the nilradical are identical to the members of the descending central chain of the associated Lie algebra. Thus, the Lie nilpotency class is identical to the associative nilpotency class of the associative nilradical. By using this result and the theorem of Xiankun Du (see theorem 3) we derive:

Theorem 5 *Let K be a field and $n \in \mathbb{N}$. The following identities are valid:*

(i) $cl(s\delta_{o,n}) = cl(s\delta_{u,n}) = n$

(ii) $cl((s\delta_{o,n})^\circ) = cl((s\delta_{u,n})^\circ) = n$

(iii) $cl((s\delta_{o,n})^\star) = cl((s\delta_{u,n})^\star) = n$.⋄

2.3.2 Solomon algebras in characteristic zero

Let K be a field of characteristic zero and $n \in \mathbb{N}$. By D_n we denote the Solomon algebra. It is defined by the K-span of class sums of so-called defect classes of KS_n: if $\alpha \in S_n$ is valid, then we define $D(\alpha) := \{i \mid i\alpha > (i+1)\alpha\}$. Solomons algebra is the K-span of $\{\sum_{D(\alpha)=D} \alpha \mid D \subseteq \underline{n-1}\}$. The surprising insight of Louis Solomon was that the product of two defect class sums is a K-linear sum of defect class sums. The Solomon algebra is of dimension 2^{n-1}, its radical factor algebra of dimension $p(n)$ - the number of partitions of n - and is isomorphic to $K^{p(n)}$. All radical complements are self-centralizing. A deep insight in this theory is included in the dissertation of Thorsten Bauer [4] (especially in chapter 3.)

Within [77] it is proven for an associative algebra possessing a self-centralizing radical complement and a radical factor algebra isomorphic to a n-fold direct product of the base field that the associative powers of the nilradical are identical to the members of the descending central chain of the associated Lie algebra. By using this result we derive by using the theorem of Xiankun Du (see theorem 3) and a result of Michael D. Atkinson for the class of nilpotency of $rad(D_n)$ (see [1], $cl(rad(D_n)) = n - 1$) for the classes of nilpotency:

Theorem 6 *Let K be a field of characteristic zero and $n \in \mathbb{N}$. The identity $cl(rad(D_n)) = cl(rad(D_n)^\circ) = cl(rad(D_n)^\star) = n - 1$ is valid.*⋄

2.3.3 Solomon-Tits algebras

Let K be a field and $n \in \mathbb{N}$. By $S(n,k)$ we denote the so-called Stirling number of $k \in \underline{n}_0$. This number is the quantity of all unordered set partitions of $\underline{n}$ possessing exactly k subsets of $\underline{n}$. In [57] and [77] the Solomon-Tits algebra $K\Pi_n$ is presented and analyzed in details by Manfred Schocker and by the author. $K\Pi_n$ is defined based on the monoid Π_n which consists of all ordered set partitions of $\underline{n}$. If $(P_1, \cdots, P_l)$ and $(Q_1, \cdots, Q_k)$ are two of these ordered partitions, then their product $\wedge_n$ is defined by

$$(P_1, \cdots, P_l) \wedge_n (Q_1, \cdots, Q_k) :=$$
$$(P_1 \cap Q_1, P_1 \cap Q_2, \cdots, P_1 \cap Q_k, \cdots, P_l \cap Q_1, P_l \cap Q_2, \cdots, P_l \cap Q_k)^\emptyset.$$

$\emptyset$ symbolizes that empty sets are deleted from this tuple. Again, $K\Pi_n$ is an example of an associative algebra possessing a factor algebra by its nilradical which is isomorphic to a r-fold product of the base field. In addition, every radical complement is self-centralizing.

The nilradical of $K\Pi_n$ is described within chapter 2 in [77]. Its dimension is $dim_K(rad(K\Pi_n)) = \sum_{k=0}^{n} (k! - 1) S(n,k)$ (see corollary 8 in [77]). The factor algebra by the nilradical is of dimension $B(n)$ – the so-called nth Bell number.

Again, by results in [77] the associative powers of the nilradical are identical to the members of the descending central chain of the associated Lie algebra. Manfred Schocker (see [57]) has proven that the associative nilpotency class of the nilradical of $K\Pi_n$ is exactly n. We conclude by using theorem 3 – the theorem of Xiankun Du:

Theorem 7 *Let K be a field and $n \in \mathbb{N}$. The following identity is valid:* $cl(rad(K\Pi_n)) = cl(rad(K\Pi_n)^\circ) = cl(rad(K\Pi_n)^\star) = n.\diamond$

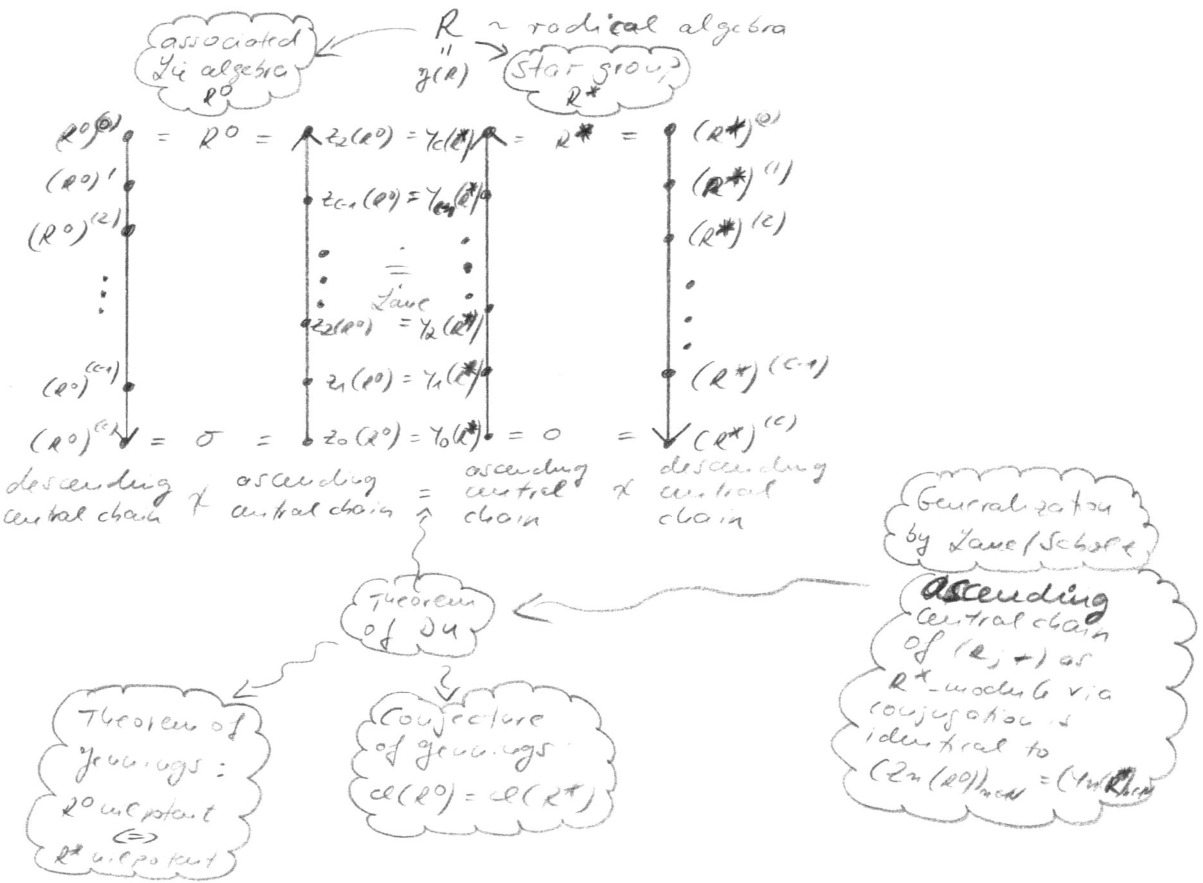
R ~ radical algebra
associated Lie algebra R°
star group R*
descending central chain
ascending central chain
Theorem of Du
Conjecture of Jennings
Theorem of Jennings: R° nilpotent ⇔ R* nilpotent
Generalization by Laue/Scholz
Ascending central chain of (R, +) as R*-module via conjugation is identical to
Laue

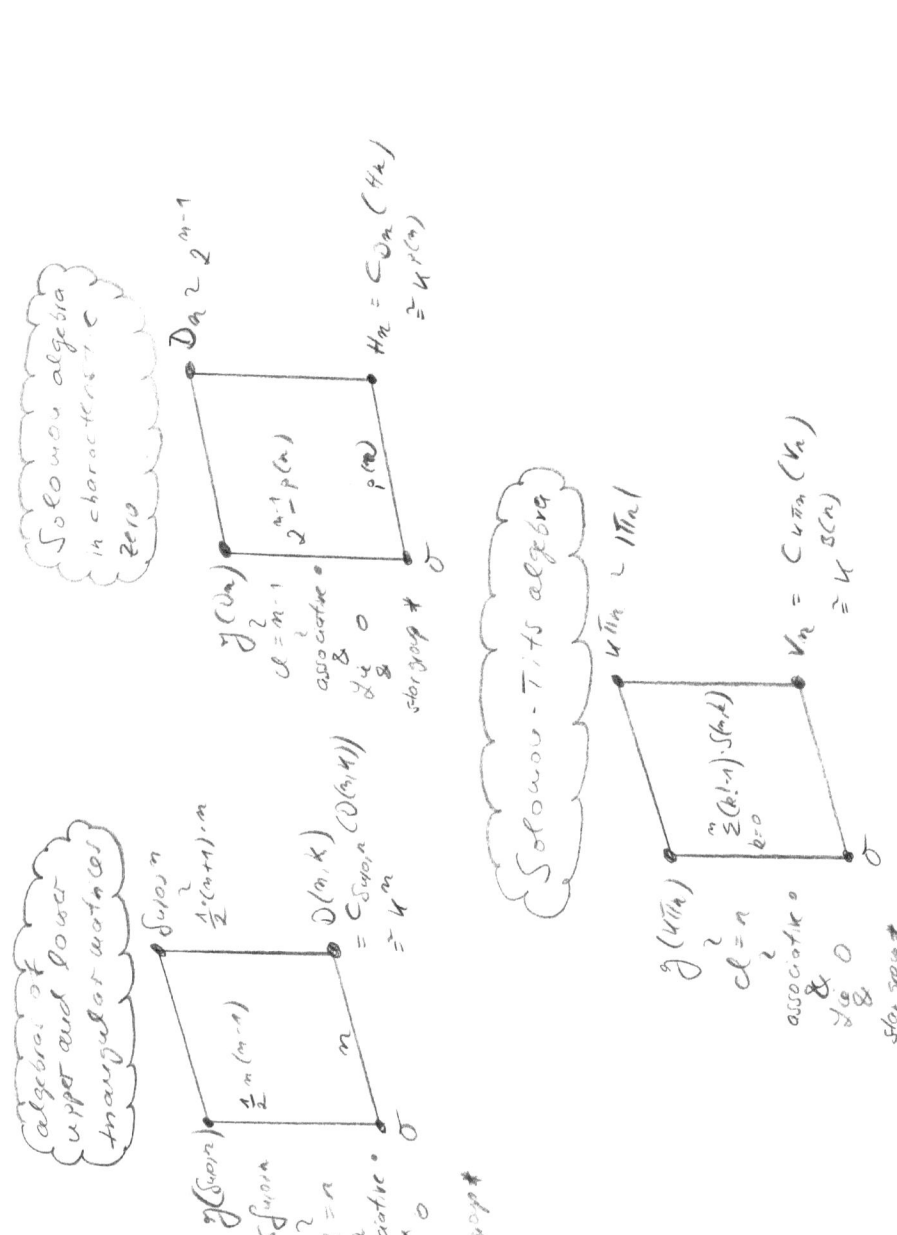

2.4 Open-ended questions and exercises

Open-ended questions 1 *(i) Does a pendant of the theorem of Du exist for radical algebras concerning solvability?*

(ii) Is it true that the factor groups along the descending central chain of the quasi regular group of a radical algebra are elementary-p-abelian if the base field is of characteristic $p > 0$ (except for the derived subgroup itself)?

(iii) Determine the order of the elementary-p-abelian factor groups along the ascending central chain of the quasi-regular group of a radical algebra if the base field is of characteristic $p > 0$ (except for the center itself)!

(iv) What is the answer for the previous questions for nilradicals of group algebras? This question is partly answered by the dissertation of M. Theede (see [69]).

(v) Determine those nilpotent algebras for which the sets of members of the series of upper and lower Lie central chains are identical.

(vi) Determine those nilpotent algebras for which the lower Lie central chain and the associative powers are identical.

(vii) Determine those nilpotent algebras for which the sets of members of the series of upper Lie central chain and associative powers are identical.

Excercise 2 *For the algebras of strict upper and lower triangular matrices over a field analyze the connections between the upper and lower Lie central chain as well as the associative powers! (Tip: see [38])*

Excercise 3 *Prove proposition 1 in details!*

Excercise 4 *Prove proposition 2!*

Excercise 5 *Let K be a field and $n \in \mathbb{N}$. Summarize the classes of nilpotency within theorem 5 for $n \leq 20$. In addition, calculate the dimension of the algebra, the nilradical and the factor algebra by the nilradical for $n \leq 20$. Are these numbers convergent for $n \to \infty$?*

Excercise 6 *Let K be a field and $n \in \mathbb{N}$. Summarize the classes of nilpotency within theorem 7 for $n \leq 20$. In addition, calculate the dimension of the algebra, the nilradical and the factor algebra by the nilradical for $n \leq 20$. Are these numbers convergent for $n \to \infty$?*

Excercise 7 Let K be a field, $char(K) = 0$ and $n \in \mathbb{N}$. Summarize the classes of nilpotency within theorem 6 for $n \leq 20$. In addition, calculate the dimension of the algebra, the nilradical and the factor algebra by the nilradical for $n \leq 20$. Are these numbers convergent for $n \to \infty$?

Excercise 8 Let A be an associative K-algebra and $c \in K$. We define $a \star_c b := a + b + cab$. A is – equipped with the composition $\star_c$ – a monoid possessing the neutral element 0. If c, d are units in $(K; \cdot)$, then the monoids $(A; \star_c)$ and $(A; \star_d)$ are isomorphic. Are these monoids isomorphic to $(A; \star)$? If $A = K$ is a finite field, then find two examples of c, d such that the monoids $(A; \star_c)$ and $(A; \star_d)$ are not isomorphic.

Excercise 9 If A is an associative unitary K-algebra, then the monoids $(A; \cdot)$ and $(A; \star)$ are isomorphic. What is the consequence for their groups of units?

Excercise 10 Determine $Q(A)$ for the following cases of an associative K-algebra A:

(i) $A = \mathbb{Z}$

(ii) A is a field.

(iii) A is a division algebra.

(iv) $A = K^{2 \times 2}$

(v) A is nilpotent.

(vi) A is nil.

(vii) A is a radical algebra.

(viii) The inverse algebra of A.

(ix) A is a direct product of algebras.

(x) A is unitary.

(xi) A is semidirect decomposition of the nilradical and a radical complement.

Which of these algebras are radical algebras? On what terms are these algebras radical algebras?

Excercise 11 Let A be an associative K-algebra and $z \in A$. On what terms are the sets $\{x - xz \mid x \in A\}$ and A identical? Answer the same question for the set $\{x - zx \mid x \in A\}$!

Excercise 12 Let A be a radical K-algebra and c an unit of K. Prove for all $n \in \mathbb{N}_0$ the identity $Y_n((A;\star)) = Y_n((A;\star_c))$. (Tip: theorem of Xiankun Du and exercise 8)

Excercise 13 Let K be a field and G a finite group. The center of KG is K-linear generated by the conjugacy class sums of G.

Excercise 14 Let p be a prime number, K a field of characteristic p and P a p-group. Analyze the following statements:

(i) The nilradical $rad(KP)$ is the augmentation ideal $Aug(KP) = \langle\{g-1 \mid g \in P\}\rangle_K$. (Tip: The elements $g - 1$ are nilpotent and generate an ideal. Apply a theorem of Joseph Wedderburn to this result!)

(ii) $K1_G$ is a central radical complement.

(iii) $(KG)^\circ$ is nilpotent, and its class of nilpotency is identical to the one of $rad(KG)^\circ$.

(iv) $E(KG)$ is the direct product of $1 + rad(KG)$ and $(K \setminus \{0\})1_G$.

(v) What is the quasi-regular group of KG?

(vi) G is contained in $1 + rad(KG)$.

(vii) If K is finite, then determine the order of $1 + rad(KG)$.

(viii) Let K be finite. Focus on the factor groups along the ascending central chain of $1+rad(KG)$ and describe the structure of these factor groups.

(ix) In what way is it possible to apply the theorem of Xiankun Du to the previous part?

Excercise 15 Let K be a finite field of characteristic 2 and $G = Q_8$ or $G = D_8$. Determine the ascending central chain of $rad(KG)^\circ$, calculate the class of nilpotency of $1+rad(KG)$ and determine the structure of the center and the factor groups along the ascending central chains of $1 + rad(KG)$. (Tip: exercise 14, theorem of Xiankun Du and theorem 4)

Excercise 16 Prove the following statement:

For all $i \in \underline{p^n - 1}$, the prime number $p = char(K)$ is a divisor of $\binom{p^n}{i}$.

What is the importance of this result within this chapter?

Excercise 17 Let A, B be associative radical algebras. Prove that $A \times B$ is a radical algebra. Apply the theorem of Xiankun Du to $A \times B$ and determine the ascending central chain and class of nilpotency for $(A \times B)^\circ$. Is the latter Lie algebra identical to $(A^\circ) \times (B^\circ)$? Analyze star groups of direct products of radical algebras, too.

Excercise 18 *(eAe)* Let A be an associative K-algebra and e an idempotent of A. If A is a radical algebra, then eAe is a radical algebra. (Tip: Do a research in the literature and determine the nilradical of eAe!)

Excercise 19 *(zero-extension)* Let A be a K-algebra based on a composition $\cdot$. On the K-space $B := A \times A$ a multiplication $\odot$ is defined by $(a,x) \odot (b,y) := (ab, ay+xb)$. True or false: B is a radical algebra if and only if A is a radical algebra.

Excercise 20 Let K be a field. Determine the ascending central chain of $(s\delta_{u,3})^\circ$ and $(s\delta_{u,4})^\circ$. For K being finite of characteristic $p > 0$ determine the structure of the center and the factor groups along the ascending central chains of these algebras. Is there a conjecture for arbitrary n?

Excercise 21 Let K be a field. Determine the ascending central chain of $(s\delta_{o,3})^\circ$ and $(s\delta_{o,4})^\circ$. For K being finite of characteristic $p > 0$ determine the structure of the center and the factor groups along the ascending central chains of these algebras. Is there a conjecture for arbitrary n?

Excercise 22 Determine the defect classes of S_3 and S_4! Calculate all binary products of defect classes (also the squares). Is it possible to represent each product as a sum of defect class sums possessing coefficients in $\mathbb{N}$, $\mathbb{Z}$ or $\mathbb{Q}$?

Excercise 23 Let A be an associative nilpotent K-algebra. A° and $A^\star$ are nilpotent, and the identity $cl(A) \geq cl(A^\circ) = cl(A^\star)$ is valid. (Tip: bound the powers of A by the members of the descending central chain, use the theorem of Xiankun Du)

Excercise 24 Let K be a field and A a finite-dimensional associative K-algebra possessing a self-centralizing radical complement which is isomorphic to a n-fold product of the base field K. The identity $cl(rad(A)) = cl(rad(A)^\circ) = cl(rad(A)^\star)$ is valid. (Tip: [77], calculate the descending central chain, theorem of Xiankun Du)

Excercise 25 Let A be an associative K-algebra, $x_1, \ldots, x_n \in A$ and $n \in \mathbb{N}$. Calculate $x_1 \star \cdots \star x_n$ for $n \leq 3$. For the algebra of real quaternions determine $i \star j$ and $i \star j \star k$.

Excercise 26 Within exercise 25 calculate $\underbrace{x_1 \star \cdots \star x_1}_{n-fold}$. What is the result of this n-fold star product within the quaternion algebra for $x_1 \in \{i,j,k\}$? Is the calculation dependent on the base field K?

Excercise 27 Let $K := GF(3)$. Within $K^{2\times 2}$ determine two non-commuting matrices A, B and calculate $A \circ B$, $A \circ B \circ B$ and $A \circ B \circ B \circ B$. Calculate the same products by switching A and B. What are the results for $GF(2)$?

Excercise 28 *What results can be deduced, if within exercise 27 the composition ∘ is replaced by ∗?*

Excercise 29 *What results can be derived by corollary 1, if the composition ⋆ is replaced by $\star_c$? What is the answer if c is an unit (see exercise 8)?*

Excercise 30 *If A is an associative algebra, then every nilpotent element is quasi-regular.*

Excercise 31 *Let A be an associative K-algebra. A can be extended to an unitary associative K-algebra A^K (see e.g. series I or [75] for an exact definition and construction of A^K). Is there a connection between the quasi regular groups of A and A^K and the group of units of A^K? Is A a radical algebra if and only if A^K is a radical algebra?*

Excercise 32 *Prove that for a finite-dimensional associative unitary algebra every element is an unit or a zero divisor (see [77]). What is the definition of a zero divisor?*

Excercise 33 *By using exercise 31 try to transfer the results of exercise 32 to non-unitary algebras.*

Excercise 34 *Let p be a prime number, K be a finite field of characteristic p and A be an associative finite-dimensional K-algebra. Let us focus on the associative powers of the nilradical $rad(A)$ which is the series $(rad(A)^n)_{n\in\mathbb{N}}$. Prove the following statements*

(i) There are only finite many associative powers of the radical.

(ii) For all $i \in \mathbb{N}$ the factor algebra $rad(A)^i/rad(A)^{i+1}$ is a zero-algebra.

(iii) For all $i \in \mathbb{N}$ the group $(rad(A)^i/rad(A)^{i+1})^\star$ is elementary-p-abelian.

Excercise 35 *In view of exercise 34 find the decomposition into cyclic groups of order p for the star groups $(rad(A)^i/rad(A)^{i+1})^\star$ for all $i \in \mathbb{N}$ with respect to the following algebras based on a finite field of characteristic $p > 0$:*

(i) The algebra of upper triangular matrices.

(ii) The algebra of lower triangular matrices.

(iii) The Solomon-Tits algebra (see the article of Manfred Schocker [57] for the associative powers of the nilradical).

(iv) Let G be a cyclic p-group generated by z. Then KG decomposes into $rad(KG)$ and $K1_G$. The radical is exactly $KG(1-z)$ (see e.g. [76]).

Excercise 36 *Prove theorem 4 in details!*

Excercise 37 *Why is within the proof of theorem 4 the statement not valid for the center itself?*

Chapter 3

Solvability

Within this chapter we analyze for a finite-dimensional associative unitary K-algebra the connection of solvability between the associative algebra, its group of units and their associated Lie algebra. We will prove that for adequate fields these three conditions are equivalent.

An analogue to the theorem of Xiankun Du for the class of solvability is not known by the author. But within our examples a close connection for the class of solvability between the associative algebra, its group of units and their associated Lie algebra is obtained.

3.1 Solvability of the associated Lie algebra

Definition and remark 2 Let A be an associative K-algebra. A is called solvable, if $A/rad(A)$ is commutative.

For a Lie algebra L resp. for a group G let $(L^{[n]})_{n \in \mathbb{N}}$ resp. $(G^{[n]})_{n \in \mathbb{N}}$ the descending sequence of derived subalgebras resp. subgroups (also called commutator subalgebras resp. subgroups) of L resp. G. L resp. G is called solvable, if the sequence of derivations is reaching the trivial subalgebra resp. subgroup after finite many steps. The minimal number of these steps is called the class of solvability – denoted by $st(L)$ resp. $st(G)$.

For an associative algebra A the class of solvability can be defined, too. For this, we focus on the ideal of A generated by $A \circ A$. This ideal is called the derived subalgebra of A and is denoted by $A^{[1]}$ or by A'. For every $n \in \mathbb{N}$ we define the nth derived subalgebra or commutator ideal of A by $A^{[n+1]} := (A^{[n]})'$. A is solvable as associative algebra if and only if an element $m \in \mathbb{N}$ exists such that the ideal of A generated by $A^{[m]}$ is zero. The minimal m possessing this condition is called the class of solvability of A – denoted by $st(A)$. It is straightforward to prove that for A being solvable the Lie algebra A° is solvable, too, and that $st(A) \geq st(A^\circ)$ is valid. ⋄

Proposition 3 *If A is an associative right artian solvable K-algebra, then A° is solvable.*

Proof. A is solvable, and hence $A^\circ \circ A^\circ \subseteq rad(A)$ is valid. By using the associative nilpotency of $rad(A)$ we conclude the Lie nilpotency of $rad(A)^\circ$. $rad(A)^\circ$ is a subalgebra of A°, and we conclude that $A^\circ \circ A^\circ$ is nilpotent. Hence A° is solvable. ⋄

To prove the opposition implication of proposition 3 we use a well-known strategy within the theory of associative algebras: we begin the proof for central division algebras, afterwards for division algebras, then for simple and semi-simple algebras and at the end for an arbitrary algebra. The next two remarks are related to this strategy.

Remark 1 *Let K be a field and A, B associative unitary K-algebras. The following statements are valid:*

(i) For all $a, c \in A$ and $b, d \in B$: $(a \otimes b) \circ (c \otimes d) = (a \circ c) \otimes bd + ca \otimes (b \circ d)$.

(ii) If A° is solvable and B° is abelian, then $(A \otimes_K B)^\circ$ is solvable.

Proof. ad(i): This can be verified by a straightforward calculation.

ad(ii): B° is abelian, and thus for all $x, y \in A$ and $c, d \in B$ we derive by (i): $(*)$ $(x \otimes c) \circ (y \otimes d) = (x \circ y) \otimes (cd)$.
Let $T := A \otimes_K B$. By using $(*)$ and an induction argument we conclude for all $m \in \mathbb{N}$ that the identity $(T^\circ)^{[m]} \subseteq (A^\circ)^{[m]} \otimes_K B$ is valid. A° is solvable, and thus part (ii) is proven by using the solvability of A°. ⋄

Within the second remark we analyze full matrix algebras concerning solvability.

Example 1 (i) Let K be a field. $A := K^{2 \times 2}$ is a simple associative K-algebra. We prove that A° is solvable if and only if $char(K) = 2$ is valid. For this, let $B := \{e_{11}, e_{12}, e_{21}, e_{22}\}$ be the standard basis of A. (e_{ij} is defined to be the matrix such that only the $(i; j)$-entry is different from zero and in addition equal to 1.) It is straightforward to calculate the identities $e_{22} \circ e_{21} = e_{21}$, $e_{22} \circ e_{12} = -e_{12}$, $e_{22} \circ e_{11} = 0_A$, $e_{21} \circ e_{12} = e_{22} - e_{11}$, $e_{21} \circ e_{11} = e_{21}$ and $e_{12} \circ e_{11} = -e_{12}$. Hence, $(A^\circ)^{[1]} = \langle e_{21}, e_{12}, e_{22} - e_{11} \rangle_K$ is valid. In addition, $e_{21} \circ e_{12} = e_{22} - e_{11}$, $e_{21} \circ (e_{22} - e_{11}) = -2_K e_{21}$ and $e_{12} \circ (e_{22} - e_{11}) = -2_K e_{12}$ are valid. If $char(K) = 2$ is assumed, then $(A^\circ)^{[2]} = \langle e_{22} - e_{11} \rangle_K$ and $(A^\circ)^{[3]} = \{0_A\}$ (and thus $st(A^\circ) = 3$) are true. In the other case $(A^\circ)^{[1]} = (A^\circ)^{[2]}$ is valid and A° is not solvable.

(ii) Let K be a field and $n \in \mathbb{N}_{\geq 3}$. We prove that $(K^{n \times n})^\circ$ is not solvable. Let $\{e_{ij} \mid 1 \leq i,j \leq n\}$ the standard basis of $K^{n \times n}$ as used in (i). It is sufficient to prove that $(K^{3 \times 3})^\circ$ is not solvable because $(K^{n \times n})^\circ$ is containing a subalgebra isomorphic to $(K^{3 \times 3})^\circ$. The identities $e_{11} \circ e_{12} = e_{12}$, $e_{11} \circ e_{13} = e_{13}$, $e_{11} \circ e_{31} = -e_{31}$, $e_{21} \circ e_{22} = e_{21}$, $e_{23} \circ e_{33} = e_{23}$, $e_{32} \circ e_{33} = -e_{32}$, $e_{12} \circ e_{21} = e_{11} - e_{22}$ and $e_{13} \circ e_{31} = e_{11} - e_{33}$ are valid. In addition, $e_{12} \circ e_{21} = e_{11} - e_{22}$, $e_{13} \circ e_{31} = e_{11} - e_{33}$, $e_{12} \circ e_{23} = e_{13}$, $e_{13} \circ e_{32} = e_{12}$, $e_{32} \circ e_{21} = e_{31}$, $e_{21} \circ e_{13} = e_{23}$, $e_{23} \circ e_{31} = e_{21}$ and $e_{31} \circ e_{12} = e_{32}$ are true. We derive $\langle e_{12}, e_{13}, e_{21}, e_{23}, e_{31}, e_{32}, e_{11} - e_{22}, e_{11} - e_{33}\rangle_K \subseteq ((K^{3 \times 3})^\circ)^{[n]}$ for all $n \in \mathbb{N}$. Hence, $(K^{3 \times 3})^\circ$ is not solvable.

(iii) Let K be a field, $char(K) = 2$ and A a 4-dimensional central-simple associative unitary algebra. By using [36] a K-basis $B := \{1_A, i, j, k\}$ of A and elements $a \in K$, $b \in K \setminus \{0_K\}$ exist such that $i^2 + i = a1_A$, $j^2 = b1_A$ and $ij = k = j(i + 1_A)$ are valid. These algebras are called generalized quaternion algebras in characteristic two. We prove that A° is solvable. The identities $i \circ j = ij + ji = j$, $i \circ k = ik + ki = i(ij) + (ji + j)i = (i + a1_A)j + j(i + a1_A) + ji = ij + aj + ji + aj + ji = ij = k$ and $j \circ k = jk + kj = j(ji + j) + ij^2 = j^2i + j^2 + ij^2 = bi + j^2 + bi = j^2 = b1_A$ are valid. We conclude $(A^\circ)^{[1]} = \langle j, k, b1_A\rangle_K$. Hence, $(A^\circ)^{[2]} = \langle b1_A\rangle_K$ and $(A^\circ)^{[3]} = \{0_A\}$ are true. ◇

The following lemma is the key element related to our strategy:

Lemma 1 *Let K be a field and D a central-simple and finite-dimensional associative K-division algebra. If $char(K) = 2$ is valid, then let the K-dimension of D be not 4. The following statements are equivalent:*

(i) D is solvable.

(ii) $D = K1_A$

(iii) D° is solvable.

Proof. Because of $rad(D) = \{0\}$ part (i) is equivalent to part (ii). By using proposition 3 part (ii) results in part (iii).
We assume part (iii). Let L be a maximal subfield of D and $n = dim_K(L) = dim_L(D)$. Then (see e.g. [50]) the statement $D \otimes_K L \cong_{A_1} L^{n \times n}$ is valid. Thus, $n^2 = dim_K(D)$ is true, too. Remark 1 implies that $(L^{n \times n})^\circ$ is solvable as K-algebra. If $n = 1$ is valid, then part (ii) is true. We assume $n \in \mathbb{N}_{\geq 2}$. If $n \neq 2$ is valid, then the K-algebra $L^{n \times n}$ contains a subalgebra T isomorphic to $K^{3 \times 3}$. Thus, T° is a K-subalgebra of $(L^{n \times n})^\circ$ and solvable, too. This is a contradiction to remark 1. For $n = 2$ we have to analyze only the case $char(K) \neq 2$ (due to our assumptions). The K-algebra $L^{n \times n}$

contains a subalgebra T isomorphic to $K^{2\times 2}$. T° is solvable as K-subalgebra of $(L^{n\times n})^\circ$. This is a contradiction to remark 1. ⋄

Now we are prepared to prove the connection between the solvability of an associative algebra and its associated Lie algebra.

Theorem 8 *Let K be a field, $char(K) \neq 2$ and A a finite-dimensional associative K-algebra. The following statements are equivalent:*

(i) A is solvable.

(ii) A° is solvable.

Proof. Based on proposition 3 only the implication from (ii) to (i) is to be proven. This proof is divided into several steps:

Step 1: Let A be a K-division algebra.
We use induction on $dim_K(A)$ to prove the result within this step. In dimension one the theorem is true. If $Z(A) = K1_A$ is valid, then the theorem is a consequence of lemma 1. Thus, we assume $K1_A \neq Z(A)$. $Z(A)$ is a field and A is a $Z(A)$-algebra and $dim_{Z(A)}(A) \leq dim_K(A) - 1$ is valid. A° is solvable as K-algebra, and thus it is solvable as $Z(A)$-algebra, too. By induction A is solvable as $Z(A)$-algebra, and hence A is a field.

Step 2: Let A be simple.
By our assumption an element $n \in \mathbb{N}$ and a finite-dimensional associative K-division algebra D exist such that $A \cong_{A_1} D^{n \times n}$ is valid. It is straightforward to prove that a K-subalgebra T of $D^{n\times n}$ exists which is isomorphic to D. T° is a K-subalgebra of $(D^{n\times n})^\circ$, and hence by using step 1 the algebra D is a field. If $n \neq 1$ is valid, then $D^{n\times n}$ would possess a subalgebra X isomorphic to $K^{2\times 2}$. But X is not solvable using remark 1. Hence $n = 1$ is valid, and A is a field.

Step 3: Let A be semi-simple.
By our assumption a finite direct decomposition of A exists and each member of the decomposition is a simple algebra. Every simple ideal is a subalgebra of A°, and hence each ideal is solvable as Lie-algebra, too. By using step 2 each ideal of this decomposition is commutative. Hence A is commutative, too.

Step 4: Now we focus on the general case of A.
A° is solvable, and $A/rad(A)$ is a finite-dimensional semisimple associative K-algebra. $rad(A)^\circ$ is an ideal of A°, and thus $A^\circ/rad(A)^\circ = (A/rad(A))^\circ$ is solvable, too. The proof is finished by using step 3. ⋄

We apply the previous theorem to tensor product. For this we need the following proposition:

Proposition 4 *Let K be a field and A, B finite-dimensional associative K-algebras. If A, B are solvable, then $A \otimes_K B$ is solvable, too. If A, B are unitary, then the solvability of $A \otimes_K B$ implies the solvability of A and B.*

Proof. If A, B are unitary, then $A \otimes_K B$ contains subalgebras isomorphic to A resp. B. Thus, in this case the solvability of $A \otimes_K B$ implies the solvability of A and B.
Let A and B be solvable and $T := A \otimes_K B$. By using part (i) of remark 1 we derive $T \circ T \subseteq (A \circ A) \otimes_K B + A \otimes_K (B \circ B)$. A and B are solvable, and this implies $A \circ A \subseteq rad(A)$ and $B \circ B \subseteq rad(B)$. Hence $T \circ T \subseteq rad(A) \otimes_K B + A \otimes_K rad(B)$ is valid. The latter structure is a nilpotent ideal which is contained in $rad(T)$.⋄

By using this proposition we derive a generalized version of part (ii) within remark 1 regarding tensor products and solvability:

Theorem 9 *Let K be a field, $char(K) \neq 2$ and A, B associative finite-dimensional K-algebras. If A° and B° are solvable, then $(A \otimes_K B)^\circ$ is solvable.*

Proof. Let A° and B° be solvable. By using theorem 8 the associative algebras A and B are solvable. Proposition 4 lets us derive that $A \otimes_K B$ is solvable, and by using again theorem 8 the proof is finished. ⋄

Counterexample 1 Let K be a field and $char(K) = 2$. By using remark 1 the algebra $K^{2\times 2}$ is solvable as Lie algebra. The identity $K^{2\times 2} \otimes_K K^{2\times 2} \cong_{\mathcal{A}_1} K^{4\times 4}$ is valid. The latter algebra is as Lie algebra – again by using remark 1 – not solvable.⋄

3.2 Solvability of the group of units

Lemma 2 *Let K be a field and A a finite-dimensional associative unitary K-algebra. The identity $E(A)/(1_A + rad(A)) = E(A/rad(A))$ is valid.*

Proof. If $x \in E(A)$, then $x(1_A + rad(A)) = x + xrad(A) = x + rad(A) \in E(A/rad(A))$ is valid.
Let $a + rad(A) \in E(A/rad(A))$. There exists an element $b \in A$ such that $ab \in 1_A + rad(A)$ is valid. This implies $ab \in E(A)$. By using results in [77] the element a is an unit or a zero-divisor of A. In the first case we conclude $a + rad(A) = a(1_A + rad(A)) \in E(A)/1_A + rad(A)$. If a is a divisor of zero, then there exists an element $0_A \neq c \in A$ such that $ca = 0_A$ is valid. Hence $cab = 0_A$ is true. ab is an unit of A, and we conclude $c = 0_A$. This is a

contradiction.◊

In addition to this lemma we need the following theorem of William Raymond Scott presented in series I for our analysis:

Theorem 10 *(Scott) Let D be a K-division algebra. $E(D)$ is solvable if and only if D is a field.*◊

Now we are able to characterize associative algebras possessing solvable group of units:

Theorem 11 *Let K be a field, $\mid K \mid > 3$ and A a finite-dimensional associative unitary K-algebra. If $E(A)$ is solvable, then A is solvable.*

Add-on: The opposite implication is true without any assumptions for the field.

Proof. Let $E(A)$ be solvable. Hence $E(A)/(1_A + rad(A))$ is solvable, too. By using lemma 2 the unit group $E(A/rad(A))$ is solvable. We apply the classical result of Joseph Wedderburn and Emil Artin about the structure of associative algebra: there exist finite-dimensional unitary K-division algebras $D_1, ..., D_r$ and $n_1, ..., n_r \in \mathbb{N}$ such that $A/rad(A) \cong_{\mathcal{A}_1} \bigoplus_{i=1}^{r} D_i^{n_i \times n_i}$ is valid. As a consequence, for every $i \in \underline{r}$ there exist a subalgebra of $A/rad(A)$ which is $\mathcal{A}_1$-isomorphic to $K^{n_i \times n_i}$. We conclude that for every $i \in \underline{r}$ an isomorphic copy of $GL(n_i, K)$ is contained in $E(A/rad(A))$. These copies are solvable, and by using 6.10 in [24] we derive that for all $i \in \underline{r}$ the identity $n_i = 1$ is valid (Here we use $\mid K \mid \geq 3$, too.). Hence $A/rad(A)$ is isomorphic to a direct sum of division algebras. All group of units of these division algebras are solvable, and by theorem 10 of William Raymond Scott they are all abelian. As a consequence, A is solvable.

If A is solvable, then $A/rad(A)$ is commutative. Hence the group of units of this factor algebra is abelian. This group is – by using lemma 2 – exactly $E(A)/(1_A + rad(A))$. The associative nilpotency of $rad(A)$ implies the nilpotency of the normal subgroup $1_A + rad(A)$. We conclude that $E(A)$ is solvable.◊

If we combine this result with theorem 8 we derive the following main result connecting the solvability of an associative algebra, its associated Lie algebra and its group of units:

Main theorem 2 *Let K be a field possessing at least 5 elements, $char(K) \neq 2$ and A a finite-dimensional associative unitary K-algebra. The following statements are equivalent:*

(i) A is solvable.

(ii) $A°$ *is solvable.*

(iii) $E(A)$ *is solvable.*⋄

3.3 The theorem of Sophus Lie and Borel subalgebras

Examples of solvable associative algebras will be presented is a separate section within this chapter. In every associative algebra there are solvable subalgebras (e.g. the semidirect sum of a torus and of the nilradical). One special solvable associative subalgebra is the so-called solvable radical: the sum of all solvable ideals is solvable. Thus, within finite-dimensional associative algebras there is always a biggest solvable ideal – the solvable radical. Within the theory of Lie algebra so-called Borel subalgebras play an important role: they are maximal solvable Lie subalgebras. They contain the solvable radical and they are closely connected to the Cartan subalgebras.[1] In this section we will prove that Borel subalgebras of an associated Lie algebra based on an associative finite-dimensional algebra are associative subalgebras, if the base field is of characteristic zero.

Theorem 12 *(Sophus Lie) Let K be an algebraically closed field of characteristic zero and L a finite-dimensional solvable K-Lie algebra. If V is an irreducible L-module, then V is one-dimensional.*

---

[1] Armand Borel (born 21st of May 1923 in La Chaux-de-Fonds, Switzerland; died 11th of August 2003 in Princeton, USA) was a permanent professor at the Institute for Advanced Study in Princeton, New Jersey, United States from 1957 to 1993. He worked in algebraic topology, in the theory of Lie groups, and was one of the creators of the contemporary theory of linear algebraic groups. He studied at the ETH Zürich, where he came under the influence of the topologist Heinz Hopf and Lie-group theorist Eduard Stiefel. He was in Paris from 1949: he applied the Leray spectral sequence to the topology of Lie groups and their classifying spaces, under the influence of Jean Leray and Henri Cartan. He collaborated with Jacques Tits in fundamental work on algebraic groups, and with Harish-Chandra on their arithmetic subgroups. In an algebraic group G a Borel subgroup H is one minimal with respect to the property that the homogeneous space G/H is a projective variety. For example, if G is GLn then we can take H to be the subgroup of upper triangular matrices. In this case it turns out that H is a maximal solvable subgroup, and that the parabolic subgroups P between H and G have a combinatorial structure (in this case the homogeneous spaces G/P are the various flag manifolds). Both those aspects generalize, and play a central role in the theory. The Borel-Moore homology theory applies to general locally compact spaces, and is closely related to sheaf theory. He published a number of books, including a work on the history of Lie groups. In 1978 he received the Brouwer Medal and in 1992 he was awarded the Balzan Prize "For his fundamental contributions to the theory of Lie groups, algebraic groups and arithmetic groups, and for his indefatigable action in favor of high quality in mathematical research and the propagation of new ideas" (motivation of the Balzan General Prize Committee). He died in Princeton. He used to answer the question of whether he was related to Émile Borel alternately by saying he was a nephew, and no relation.

In particular, L is isomorphic to a Lie subalgebra of the Lie algebra of upper triangular matrices over K.⋄

The theorem of Sophus Lie implies that under its assumption the Lie algebra of upper triangular matrices is the mother of all solvable Lie algebras. If we focus on the associative span of a Lie algebra contained in the algebra of upper triangular matrices, then it is straightforward to prove that it is solvable as Lie algebra (and as associative algebra). This phenomenon is true in a wider context as proven by Hartmut Laue in [42]:

Corollary 1 *(H. Laue, theorem of the associative span) Let A be a finite-dimensional associative algebra based on a field K of characteristic 0 and L be a solvable Lie subalgebra of A°. The associative subalgebra $\langle L \rangle_A$ spanned by L is solvable.*
In particular, $\langle L \rangle_A$ is solvable as Lie algebra.⋄

Theorem 1 implies that under its assumptions maximal solvable Lie subalgebras L coincide with their associative span $\langle L \rangle_A$. Thus, we conclude the following corollary:

Corollary 2 *Let A be a finite-dimensional associative (unitary) algebra over a field K of characteristic 0 and B a Borel subalgebra of A°. B is an associative (unital) subalgebra of A.*⋄

Main theorem 2 implies that the group of units of Borel subalgebra are solvable subgroups. Solvable unitary K-algebras possessing a separable factor algebra by its nilradical based on a sufficient large field K are K-linear spanned by its group of units (see e.g. remark 2). By using this and corollary 2 we conclude:

Corollary 3 *Let A be a finite-dimensional associative (unitary) algebra over a field K of characteristic 0, $\mathcal{B}_A$ be the set of Borel subalgebras of A° and $\mathcal{SU}_{E(A)}$ the set of solvable subgroups of $E(A)$. The function*

$$\mathcal{B}_A \longrightarrow \mathcal{SU}_{E(A)}, B \mapsto E(B)$$

is injective.⋄

An interesting open-ended question is whether the group of units of Borel subalgebras are maximal solvable subgroups and all such subgroups – so-called Borel subgroups – are presentable in this way. This statement would be provable if the K-linear span of a solvable subgroup is a solvable associative algebra. But this is wrong which is the content of the following example. (Later on we will proof within solvable associative algebras such a correspondence for maximal nilpotent substructures!)

Example 2 Within the real quaternion algebra $\mathbb{H}$ the quaternion group is a solvable subgroup. Its $\mathbb{R}$-linear span is $\mathbb{H}$. The quaternion algebra $\mathbb{H}$ is as associative algebra not solvable because it is no field.⋄

We finish this section by providing an example that corollary 1 is wrong in characteristic different to zero. This example is to be enhance by the reader within the exercises to an arbitrary characteristic.

Example 3 Let K be a field of characteristic 3, X, Y matrices of $K^{3\times 3}$ defined by

$$X := \begin{pmatrix} 0 & 1 & 0 \\ 0 & 0 & 1 \\ 1 & 0 & 0 \end{pmatrix} \text{ and } Y := \begin{pmatrix} 0 & 0 & 0 \\ 0 & 1 & 0 \\ 0 & 0 & 2 \end{pmatrix}.$$

Because of $3 = 0$ the identity $X \circ Y = X$ is valid. Hence $L := \langle X, Y \rangle_K$ is a solvable Lie subalgebra of $gl(3, K)$. We will prove that its associative span is not solvable as Lie algebra. For $char(K) \neq 2$ theorem 8 ensures that this statement is equivalent to the non-solvability of the associative span as associative algebra. It is straightforward to calculate (by using $3 = 0$) that the Lie product $X^2 \circ (X^2 \circ Y)$ is the unit matrix. Therefor the smallest ideal of $\langle L \rangle_A$ containing all Lie products of $\langle L \rangle_A \circ \langle L \rangle_A$ is unital. Hence the associative derivation of $\langle L \rangle_A$ is exactly $\langle L \rangle_A$ itself, and we conclude that $\langle L \rangle_A$ is not solvable.⋄

3.4 Standard examples

3.4.1 Group algebras

Let K be a field and G a finite group. In [47] it is proven that $(KG)^\circ$ is solvable if and only G is abelian or the derived subgroup of G is a p-group (in the case $p := char(K) \notin \{0, 2\}$, G non-abelian) or G possesses a subgroup U of index 2 such that the derived subgroup of U is a 2-group (in the case $char(K) = 2$, G non-abelian).

Four years later the authors in [48] analyzed on what terms $E(KG)$ is solvable: G is abelian or $G/O_p(G)$ – which is the largest normal p-subgroup of G – is abelian (in the case $p := char(K) \neq 0$, G non-abelian). By using a theorem of Sylow G' is a p-group if and only if $G/O_p(G)$ is abelian. Hence for $char(K) \neq 2$ the Lie algebra $(KG)^\circ$ is solvable if and only if $E(KG)$ is solvable. This statement is generalized by 5.4 in [4] and theorem 8. In the case $char(K) = 2$ the solvability of $E(KG)$ implies the one of $(KG)^\circ$ (see theorem 13), but the group S_3 demonstrates that $(KG)^\circ$ is but $E(KG)$ is not solvable. Now we answer the question on what terms KG is solvable.

Theorem 13 *KG is solvable if and only if $E(KG)$ is solvable.*

Proof. In the case $char(K) = 0$ the result is a consequence of theorem 5.4 in [4]. Let $p := char(K) \neq 0$. If KG is solvable, then $E(KG)$ is solvable by theorem 5.4 in [4]. Let $E(KG)$ be solvable. If G is abelian, then KG is commutative and hence solvable. Let G be not abelian. As described within the introduction to remark 4 the subgroup G' is a p-group. By using proposition 1.2 in [32] we derive the identity $KG/(KG\,Aug(KG')) \cong_{\mathcal{A}_1} K(G/G')$. Hence $KG/(KG\,Aug(KG'))$ is commutative. We have to prove that $KG\,Aug(KG')$ is nilpotent. Corollary 4.7 in [50] implies that $Aug(KG')$ is nilpotent. $KG\,Aug(KG') = Aug(KG')\,KG$ is valid (see e.g. proposition 1.2 in [32]), and thus $KG\,Aug(KG')$ is nilpotent, too.⋄

The calculation of the solvable class for solvable group algebras – viewed as solvable associative or Lie algebra as well for the group of units – is an open-ended question. Aner Shalev has proven some results related to this topic (see e.g. '[60] and [61]). Within our next examples these classes of solvability are closely connected.

3.4.2 The algebras of upper and lower triangular matrices

Let K be a field and $n \in \mathbb{N}$. In [75] the solvable classes for $A = \delta_{o,n}$ resp. $A = \delta_{u,n}$ are calculated. For all $n \in \mathbb{N}$ the identity $(A^\circ)^{[n]} = A^{[n]}) = rad(A)^{2^{n-1}}$ is valid. In particular, for the minimal m – for which $2^{m-1} \geq cl(rad(A)) = n$ is true – the identity $st(A) = st(A^\circ) = st(E(A)) = m$ is valid.⋄

3.4.3 Solomon algebras in characteristic zero

Let K be a field of characteristic zero and $n \in \mathbb{N}$. In [77] the following solvable classes are calculated: $st(D_n) = st((D_n)^\circ) = 1 + min\{l \in \mathbb{N} \mid 2^l \geq n-1\}$. The solvable class of the group of units of D_n is identical to $st(D_n)$.⋄

3.4.4 Solomon-Tits algebras

Let K be a field and $n \in \mathbb{N}$. In [77] the following solvable classes are determined: $st(K\Pi_n) = st((K\Pi_n)^\circ) = 1 + min\{l \in \mathbb{N} \mid 2^l \geq n\}$. K possesses at least three elements. The solvable class of the group of units of $K\Pi_n$ is identical to $st(K\Pi_n) = st((K\Pi_n)^\circ) = 1 + min\{l \in \mathbb{N} \mid 2^l \geq n\}$. The solvable class of the group of normalized units of $K\Pi_n$ is identical to $st(rad(K\Pi_n)) = st(rad(K\Pi_n)^\circ) = min\{l \in \mathbb{N} \mid 2^l \geq n\}$.⋄

n	$cl(rad(K\Pi_n))$	$cl(rad(K\Pi_n)^\circ)$	$st(rad(K\Pi_n))$	$st(rad(K\Pi_n)^\circ)$	$st(K\Pi_n)$	$st((K\Pi_n)^\circ)$
1	1	1	1	1	2	2
2	2	2	1	1	2	2
3	3	3	2	2	3	3
4	4	4	2	2	3	3
5	5	5	3	3	4	4
6	6	6	3	3	4	4
7	7	7	3	3	4	4
8	8	8	3	3	4	4
9	9	9	4	4	5	5
10	10	10	4	4	5	5
11	11	11	4	4	5	5
12	12	12	4	4	5	5
13	13	13	4	4	5	5
14	14	14	4	4	5	5
15	15	15	4	4	5	5
16	16	16	4	4	5	5

Table 3.1: nilpotency and solvable classes linked to $K\Pi_n$ and $(K\Pi_n)^\circ$

Table 3.1 summarizes nilpotency and solvable classes linked to the Solomon-Tits algebra.

n	$cl(rad(D_n))$	$cl(rad(D_n)^\circ)$	$st(rad(D_n))$	$st(rad(D_n)^\circ)$	$st(D_n)$	$st((D_n)^\circ)$
1	0	0	0	0	1	1
2	1	1	1	1	2	2
3	2	2	1	1	2	2
4	3	3	2	2	3	3
5	4	4	2	2	3	3
6	5	5	3	3	4	4
7	6	6	3	3	4	4
8	7	7	3	3	4	4
9	8	8	3	3	4	4
10	9	9	4	4	5	5
11	10	10	4	4	5	5
12	11	11	4	4	5	5
13	12	12	4	4	5	5
14	13	13	4	4	5	5
15	14	14	4	4	5	5
16	15	15	4	4	5	5
17	16	16	4	4	5	5

Table 3.2: nilpotency and solvable classes linked to D_n and $(D_n)^\circ$

Table 3.2 summarizes nilpotency and solvable classes linked to the Solomon algebra.

n	$st(1+rad(K\Pi_n))$	$st(E(K\Pi_n))$	$st(rad(K\Pi_n))$	$st(rad(K\Pi_n)^\circ)$	$st(K\Pi_n)$	$st((K\Pi_n)^\circ)$
1	1	2	1	1	2	2
2	1	2	1	1	2	2
3	2	3	2	2	3	3
4	2	3	2	2	3	3
5	3	4	3	3	4	4
6	3	4	3	3	4	4
7	3	4	3	3	4	4
8	3	4	3	3	4	4
9	4	5	4	4	5	5
10	4	5	4	4	5	5
11	4	5	4	4	5	5
12	4	5	4	4	5	5
13	4	5	4	4	5	5
14	4	5	4	4	5	5
15	4	5	4	4	5	5
16	4	5	4	4	5	5

Table 3.3: nilpotency and solvable classes linked to $E(K\Pi_n)$

Table 3.3 summarizes nilpotency and solvable classes linked to the Solomon-Tits algebra.

n	$st(1+rad(D_n))$	$st(E(D_n))$	$st(rad(D_n))$	$st(rad(D_n)^\circ)$	$st(D_n)$	$st((D_n)^\circ)$
1	0	1	0	0	1	1
2	1	2	1	1	2	2
3	1	2	1	1	2	2
4	2	3	2	2	3	3
5	2	3	2	2	3	3
6	3	4	3	3	4	4
7	3	4	3	3	4	4
8	3	4	3	3	4	4
9	3	4	3	3	4	4
10	4	5	4	4	5	5
11	4	5	4	4	5	5
12	4	5	4	4	5	5
13	4	5	4	4	5	5
14	4	5	4	4	5	5
15	4	5	4	4	5	5
16	4	5	4	4	5	5
17	4	5	4	4	5	5

Table 3.4: nilpotency and solvable classes for $E(D_n)$

Table 3.4 summarizes nilpotency and solvable classes related to the Solomon algebra.

n	$st(A) = st(A^\circ)$	$st(E(A))$	$st(E(A))$ over GF(2)
1	1	1	1
2	2	2	1
3	3	3	2
4	3	3	2
5	4	4	3
.	.	.	.
8	4	4	3
9	5	5	4
.	.	.	.
16	5	5	4
17	6	6	5

Table 3.5: nilpotency and solvable classes of $\delta_{u,n}$ and $\delta_{o,n}$

The following table 3.5 summarizes values for the solvable classes linked to the Lie algebras of lower and upper triangular matrices.

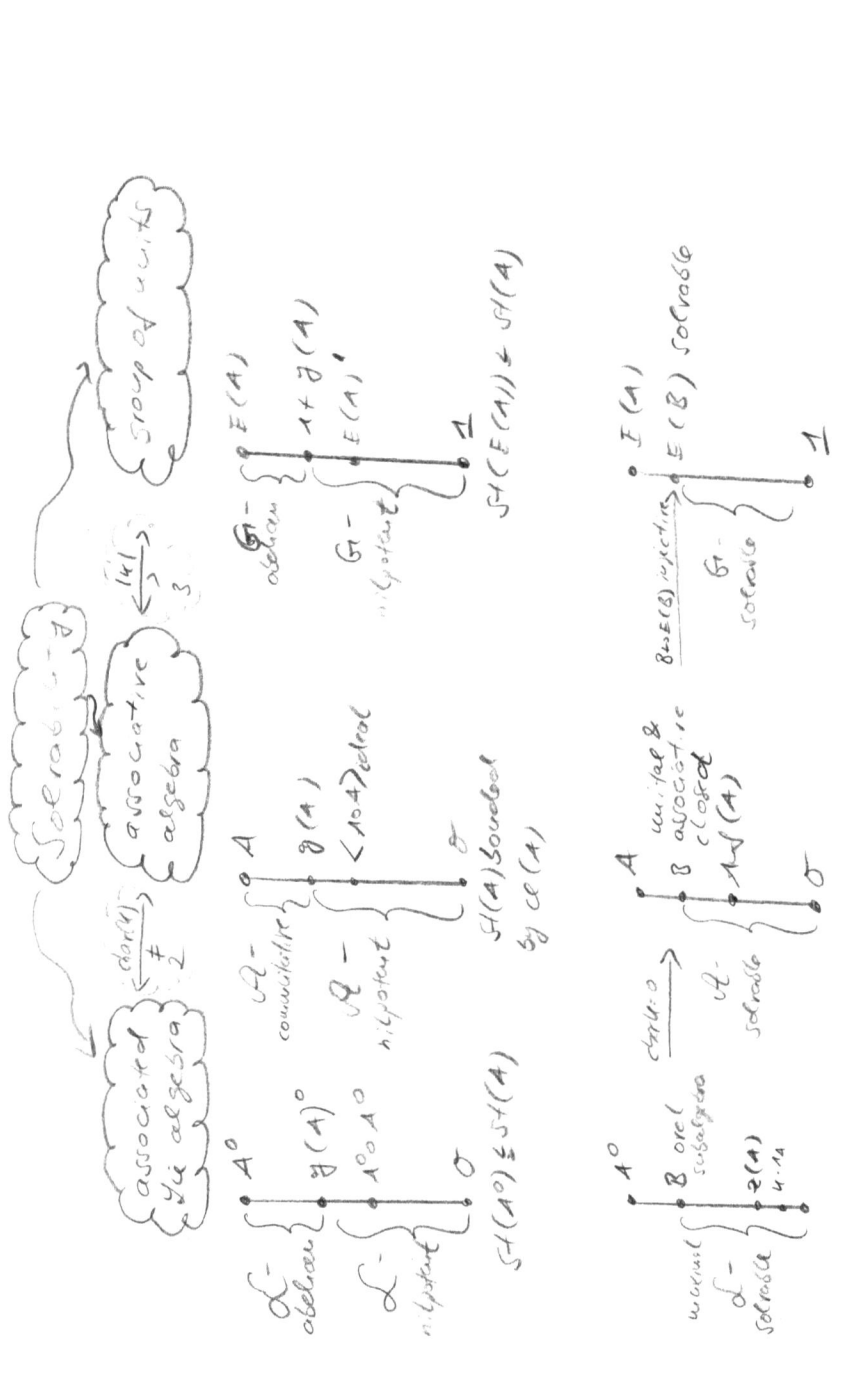

3.5 Open-ended questions and exercises

Open-ended questions 2 *(i) Let A be a finite-dimensional associative unitary solvable K-algebra and K be a field. Are there any kind of connections between the classes of solvability and of the descending chains of commutator subalgebras resp. subgroups between A, A° and $E(A)$?*

(ii) Does a special result in part (i) exist for radical algebras or nilpotent algebras?

(iii) Does a special result in part (i) exist for solvable group algebras?

(iv) Does a special result in part (i) exist for the algebra eAe for an arbitrary idempotent e (see exercise 60)?

(v) Does a description of Borel subalgebras and Borel subgroups exist within the Lie algebra and group of units associated to an associative algebra?

(vi) Determine those nilpotent algebras A for which $cl(A)$ is exactly the maximum of all nilpotency classes of the elements of A.

Excercise 38 *Determine for the radical $J(A)$ of the Solomon-Tits algebra, of the Solomon algebra in characteristic zero and of the algebras of upper and lower triangular matrices based on a field whether $cl(J(A))$ is exactly the maximum of all nilpotency classes of the elements of $J(A)$.*

Excercise 39 *Let us focus on the topic within the previous exercise 38 for special group algebras: Let G be a cyclic p-group generated by z and K be a field of characteristic p. Then KG decomposes into $rad(KG)$ and $K1_G$. The radical is exactly $KG(1-z)$ (see e.g. [76]).*

Excercise 40 *Enhance example 3 to an arbitrary prime number p and a field K of characteristic p by proving the following statements for the matrices X and Y:*

(i) $\langle X, Y \rangle_{K^\circ}$ is a solvable Lie algebra.

(ii) X and Y possess no common eigenvector.

(iii) The theorem of Lie is wrong in characteristic p.

(iv) $\langle X, Y \rangle_A$ is a non-solvable associative algebra.

(v) $\langle X, Y \rangle_A$ is a non-solvable Lie algebra equipped with the multiplication $\circ$.

Excercise 41 Analyze the quaternion algebras in characteristic $\neq 2$ on what terms the associative and Lie structure as well as the group of units are solvable. If one of these structures is solvable, then determine the class of solvability and the descending chain of commutator subalgebras resp. subgroups!

Excercise 42 Let A be an associative K-algebra. The identity $(A^\circ)/rad(A)^\circ = (A/rad(A))^\circ$ is valid. Is it possible to enhance this identity by using – instead of $rad(A)$ – an arbitrary (nilpotent) ideal I of A?

Excercise 43 Let A be an associative unitary K-algebra. The identity $E(A/rad(A)) = E(A)/(1 + rad(A))^\circ$ is valid. Is it possible to enhance this identity by using – instead of $rad(A)$ – an arbitrary (nilpotent) ideal I of A? Is it possible to enhance this identity to non-unitary associative algebras by using the star-group?

Excercise 44 True or false: Subalgebras of solvable associative algebras are solvable.

Excercise 45 True or false: Subalgebras of solvable Lie algebras are solvable.

Excercise 46 Let A be an associative unitary K-algebra and T a subalgebra of A. T is called unital, if 1_A is contained in T. T is called unitary, if T is (without keeping the focus on the substructure) an unitary K-algebra. Analyze the connections between unital and unitary subalgebras (Tip: idempotent elements!). Provide an examples of an unitary subalgebra which is not unital.

Excercise 47 Are unit groups of unital subalgebras of associative algebras possessing a solvable unit group solvable, too? Generalize this result to non-unital associative algebras!

Excercise 48 Prove the following statement in details:

> If A is solvable, then $A/rad(A)$ is commutative. Hence its group of units is abelian. This group is based on lemma 2 exactly $E(A)/(1_A + rad(A))$. $rad(A)$ is nilpotent, and hence the normal subgroup $1_A + rad(A)$ is nilpotent, too. We conclude that $E(A)$ is nilpotent.

For what reason is the statement used within this chapter?

Excercise 49 Let A be an associative unitary solvable K-algebra. A° and $E(A)$ are solvable, and the identity $max\{st(A^\circ), st(E(A))\} \leq st(A)$ is valid. Generalize this statement to non-unitary solvable algebras.

Excercise 50 Do a research within series I for the naming, definition and meaning of A^K. In what way is the composition $\star$ important for A^K?

Excercise 51 Transfer the results of main theorem 2 to non-unitary associative algebras by using exercise 50!

Excercise 52 Apply main theorem 2 to the inverse algebra of an associative algebra A! Is it possible to compare the classes of solvability for A, $E(A)$ and A° and their corresponding inverse algebra structures? Is it possible to compare the chains of descending commutator subgroups resp. subalgebras for these structures?

Excercise 53 Apply main theorem 2 to the direct product of associative algebras A and B! Is it possible to compare the classes of solvability for $A, B, E(A), E(B)$ and A°, B° and their corresponding structures of the direct product? Is it possible to compare the chains of descending commutator subgroups resp. subalgebras for these structures?

Excercise 54 Summarize the classes of solvability in section 3.4.2 for $n \leq 30$. Do these numbers converge for $n \to \infty$?

Excercise 55 Summarize the classes of solvability in section 3.4.3 for $n \leq 30$. Do these numbers converge for $n \to \infty$?

Excercise 56 Summarize the classes of solvability in section 3.4.4 for $n \leq 30$. Do these numbers converge for $n \to \infty$?

Excercise 57 On what terms are factor algebras of solvable associative or Lie algebras solvable, too?

Excercise 58 On what terms are matrix algebras over associative solvable algebras solvable, too?

Excercise 59 Analyze for the following cases of a field K and a finite group G on what terms the group algebra KG is solvable:

(i) $G = D_{2n}$

(ii) $G = Q_{4n}$

(iii) $G = SD_{2^n}$

(iv) $G = D_{2^n}$

(v) $G = Q_{2^n}$

(vi) $G = S_n$

(vii) G is abelian

(viii) $G = S_3$

(ix) $G = A_n$

(x) G is a p-group

(xi) The direct product of two groups A, B such that KA and KB are solvable.

On what terms are $E(KG)$ and $(KG)^\circ$ solvable?

Excercise 60 *(eAe)* Let K be a field, A an associative finite-dimensional unitary K-algebra possessing a separable factor algebra by its nilradical, T a radical complement of $rad(A)$ in A and e an idempotent of A. Analyze the following statements:

(i) e is diagonalizable and hence separable.

(ii) The K-subalgebra generated by $\{1, e\}$ is separable, and hence – by using an enhanced version of theorem of Wedderburn-Malcev (see e.g. [75]) – it is contained in a radical complement conjugated to T by an suitable element $1 + r$, $r \in rad(A)$.

(iii) Do a research in the literature that $e\,rad(A)\,e$ is the nilradical of eAe.

(iv) Let A be solvable. $A = rad(A) \oplus T^{1+r}$ is true, and thus $eAe = e\,rad(A)\,e \oplus eT^{1+r}e$ is valid. $eT^{1+r}e$ is a separable radical complement (Tip: Every subalgebra of a commutative separable associative algebra is separable and commutative, too (see e.g. [75]).) eAe is an associative finite-dimensional unitary K-algebra possessing a separable factor algebra by its nilradical. Determine the unit element!

(v) True or false: eAe is solvable if and only if A is solvable.

(vi) Let A and eAe be solvable. Are there any kind of connections between the classes of solvability and the descending chains of commutator subalgebras resp. subgroups between A, A° and $E(A)$ and the corresponding structures of eAe?

If the analysis is too complex, then assume that e is central.

Excercise 61 *(zero extension)* Let A be a K-algebra defined by a multiplication $\cdot$. On the K-space $B := A \times A$ we define a new multiplication $\odot$ by $(a, x) \odot (b, y) := (ab, ay + xb)$. Prove resp. analyze the following statements:

(i) $(B; \odot)$ is a K-algebra.

(ii) $(A;\cdot)$ is associative if and only if $(B;\odot)$ is associative.

(iii) $(A;\cdot)$ is commutative if and only if $(B;\odot)$ is commutative.

(iv) If $(A;\cdot)$ is unitary, then $(B;\odot)$ is unitary, and $(1_A;0)$ is the unit element.

(v) Is – within the previous part – the opposite implication true, too?

(vi) The center of $(B;\odot)$ is $(Z(A) \times Z(A); \odot)$. Is the center isomorphic to $(Z(A) \times Z(A);\cdot)$?

(vii) $0 \times A$ is an nilpotent ideal of B such that every product is zero (a so-called zero-ideal). Are the compositions $\odot$ and $\cdot$ are identical on this ideal?

(viii) $(B/(0 \times A; \odot)$ is isomorphic to $(A;\cdot)$.

(ix) If A is an associative unitary finite-dimensional separable K-algebra, then B is an associative unitary finite-dimensional K-algebra, its nilradical is $0 \times A$ and the factor algebra by the nilradical is isomorphic to $(A;\cdot)$. $A \times 0$ is a radical complement. Is this complement or is the nilradical isomorphic to $(A;\cdot)$?

(x) If A is – within the previous part – commutative, then B is solvable. What are the classes of solvability for B, B° and $E(B)$? Is it possible to describe them based on the classes of solvability of A?

(xi) Is $(B;\odot)$ isomorphic to the direct product $(A \times A;\cdot)$?

(xii) Analyze the multiplication of the inverse algebra of $(B;\odot)$.

(xiii) On what terms is B solvable?

Excercise 62 *True or false:*

(i) Every solvable associative algebra is nilpotent.

(ii) Every nilpotent associative algebra is solvable.

(iii) Every solvable associative algebra is Lie nilpotent.

(iv) Every solvable associative algebra is Lie solvable.

(v) Every Lie solvable associative algebra is solvable.

(vi) Every Lie nilpotent associative algebra is solvable.

(vii) For every nilpotent associative algebra the identity $st(A) = cl(A)$ is valid.

(viii) For every nilpotent associative algebra the identity $st(A) \leq cl(A)$ is valid.

(ix) For every nilpotent associative algebra the identity $st(A) \geq cl(A)$ is valid.

(x) For every nilpotent associative algebra the identity $st(A) \leq cl(A^\star)$ is valid.

(xi) For every nilpotent associative algebra the identity $st(A) \leq cl(A^\circ)$ is valid.

(xii) If A is an associative algebra and I is an ideal such that A/I and I are nilpotent, then A is nilpotent.

(xiii) If A is an associative algebra and I is an ideal such that A/I and I are solvable, then A is solvable.

(xiv) If A is an associative algebra and I is an ideal such that A/I and I are Lie nilpotent, then A is Lie nilpotent.

(xv) If A is an associative algebra and I is an ideal such that A/I and I are Lie solvable, then A is Lie solvable.

Excercise 63 *Let G be a finite group and N a normal subgroup of G. Prove that G is nilpotent if and only if N and G/N' are nilpotent. Analyze whether a correspondent theorem for associative or Lie algebras exist!*

Excercise 64 *Let K be a field and A, B finite-dimensional associative K-algebras. The following statements are valid:*

(i) *If $A/rad(A)$ or $B/rad(B)$ is separable, then $rad(A \otimes_K B) = rad(A) \otimes_K B + A \otimes_K rad(B)$ and $(A \otimes_K B)/rad(A \otimes_K B) \cong_{\mathcal{A}_1} (A/rad(A)) \otimes_K (B/rad(B))$ are true.*

(ii) *If $A/rad(A)$ and $B/rad(B)$ are separable, and S resp. T are radical complements in A resp. B, then $S \otimes_K T$ is a radical complement in $A \otimes_K B$. In particular, $(A \otimes_K B)/rad(A \otimes_K B)$ is separable.*

(Tip: see [75])

Chapter 4

Carter subgroups of the group of units of solvable associative algebras

In this chapter we determine the Carter subgroups of the group of units of finite-dimensional solvable unital associative algebras possessing a separable factor algebra by its nilradical. For this, we use results of the dissertation [4] of Thorsten Bauer and of the article [5] of Thorsten Bauer and Salvatore Siciliano.

We will prove a close connection between Carter subgroups and Cartan subalgebras of the associated Lie algebra. Within series I we have already determined the Cartan subalgebras, and this determination is summarized again within the next section.

4.1 The pendant within the associated Lie algebra: Cartan subalgebras

Theorem 14 *Let A be a finite-dimensional associative unitary solvable K-algebra possessing a separable factor algebra by its nilradical. The following statements are valid:*

  (i) The Cartan subalgebras of A° are exactly the centralizers of the radical complements of A. In particular, all Cartan subalgebras are unital associative K-subalgebras of A. (T. Bauer)

  (ii) All Cartan subalgebras of A° are conjugated by the normal subgroup $1_A + rad(A)$ of the group of units of A. (T. Bauer)

  (iii) The set of maximal tori of A is identical to the set of Cartan subalgebras of A° if and only if a self-centralizing radical complement exists.

(iv) The set of maximal tori of A is exactly the set of radical complements of $rad(A)$ in A.⋄

Proof. The proof was done in series I.⋄

In the next section we demonstrate that Carter subgroups are closely connected to Cartan subalgebras.

4.2 The determination of the Carter subgroups by Thorsten Bauer

We start this section by proving the following basic lemma:

Lemma 3 *(T. Bauer, S. Siciliano) Let K be a field and A an associative finite-dimensional unitary K-algebra. If H is a separable radical complement of A, then the following statements are valid:*

(i) If J is a nilpotent ideal of A, then $1 + J$ is a nilpotent normal subgroup of $E(A)$.

(ii) $E(A) = E(H) \ltimes (1 + rad(A))$

(iii) Let A be solvable. $\langle E(H) \rangle_K$ is an algebra complement of $rad(A) = rad\langle E(A)\rangle_K$ in $\langle E(A)\rangle_K$.

(iv) If A is solvable and B is a subalgebra of A, then $rad(B) = B \cap rad(A)$ is valid and $B/rad(B)$ is separable.

Proof. ad(i)+(ii): see e.g. [75].

ad(iv): see e.g. the section linked to basic algebras in series I.

ad(iii): A K-linear span of a subgroup of $E(A)$ is an associative unital subalgebra of A. $E(A)$ is containing $1 + rad(A)$, and hence the K-linear span of $E(A)$ is containing $rad(A)$. By using part (iv) the subalgebra $rad(A)$ is exactly the nilradical of the K-linear span of $E(A)$. It is straightforward to prove that $\langle E(H)\rangle_K$ is direct to $rad(A)$. Let x be an unit of A. By using the identity $x = (1 + r)h = h + hr$ we derive part (ii), and hence the proof is completed. ⋄

The next result is very important for the determination of the Carter subgroups. In addition, it is one of the key lemmas for the determination of the Fitting subgroup and for all maximal nilpotent substructures within this work. Belike, it can be of independent interest for the reader, too:

Lemma 4 *(T. Bauer, S. Siciliano) Let A be a finite-dimensional solvable unitary associative K-algebra possessing a separable factor algebra by its nilradical. If N is a nilpotent subgroup of $E(A)$, then $\langle N \rangle_K$ is a Lie-nilpotent associative subalgebra of A.*

Proof. Let $B := \langle N \rangle_K$, and we focus (see lemma 3) on a algebra complement H of $rad(B)$ in B. We want to prove $B = C_B(H)$, and by this and theorem 14 the proof is finished. Let $\bar{h} \in H$ and $g_1, g_2, \ldots, g_r \in N$, $\alpha_1, \alpha_2, \ldots, \alpha_r \in K$ such that $\bar{h} = \sum_{i=1}^{r} \alpha_i g_i$ is valid. By using lemma 3 we derive $E(B) = E(H) \ltimes (1 + rad(B))$, and thus for every $i \in \underline{r}$ there exist elements $h_i \in E(H)$ and $y_i \in rad(B)$ such that $g_i = h_i(1 + y_i)$ is valid. We conclude:

$$\bar{h} = \sum_{i=1}^{r} \alpha_i h_i (1+y_i) = \sum_{i=1}^{r} \alpha_i h_i + \sum_{i=1}^{r} \alpha_i h_i y_i,$$

and hence $\sum_{i=1}^{r} \alpha_i h_i y_i \in H \cap rad(B) = \{0\}$ is valid.

In addition, for every $i \in \underline{r}$ the identity $h_i = g_i(1+y_i)^{-1} \in N(1 + rad(B))$ is valid. Thus we conclude

$$H = \langle E(H) \cap (N(1 + rad(B))) \rangle_F. \tag{4.1}$$

Let

$$J := \{x \in B | \forall h \in H \, \exists n \in \mathbb{N} \, x(\operatorname{ad} h)^n \in rad(B)^2\}.$$

B is solvable, and hence J is an ideal of B. By using part (1) of lemma 3 we deduce that $1 + (rad(B) \cap J)$ is a normal subgroup of $E(B)$. The solvability of B implies the identity $N' \leq 1 + rad(B)$. We prove that $N' \leq 1 + (rad(B) \cap J)$ is valid, too:

Let $z \in N'$. $x := z - 1$ is an element of $rad(B)$, and thus $z^{-1} = \sum_{i \in \mathbb{N}_0} (-1)^i x^i$ is true. Let $\hat{h} \in E(H) \cap (N(1 + rad(B)))$, $g \in N$ and $y \in rad(B)$ such that $\hat{h} = g(1+y)^{-1}$ is valid. The identity

$$[z, g] = \Big(\sum_{i \in \mathbb{N}_0} (-1)^i x^i \Big) g^{-1}(1+x)g \equiv 1 + x^g - x \quad \mod rad(B)^2 \tag{4.2}$$

is true. Because of $(1+y)^{-1} = \sum_{i \in \mathbb{N}_0} (-1)^i y^i$ and (4.2) we deduce

$$[z, g] \equiv 1 + (1+y)^{-1} \hat{h}^{-1} x \hat{h} (1+y) - x \equiv 1 + \hat{h}^{-1}(x, \hat{h}) \quad \mod rad(B)^2.$$

Hence, for all $n \in \mathbb{N}$ the following statement is valid:

$$[[\ldots [z, \underbrace{g] \ldots]g}_{n} \in 1 + \hat{h}^{-n}(x(\operatorname{ad} \hat{h})^n) + rad(B)^2. \tag{4.3}$$

By using the nilpotency of N a natural number m exists such that

$$[[\ldots[z,\underbrace{g]\ldots]g}_{m}]=1$$

is true. Equation (4.3) results in $x(\operatorname{ad}\hat{h})^m \in rad(B)^2$. The elements of $E(H) \cap (N(1+rad(B)))$ commute pairwise, and thus (4.1) results in $x \in J$. Hence, $N' \subseteq 1 + (rad(B) \cap J)$ is valid.

We conclude that for all $a, a' \in N$ an element $s \in rad(B) \cap J$ exists such that $aa' = a'a(1+s)$ is valid. We deduce $[a,a'] = a'as \in rad(B) \cap J$. In particular, $[B,H] \subseteq [B,B] \subseteq J$ is true. Thus, for all $b \in B$, $h \in H$ exists a natural number n such that $b(\operatorname{ad} h)^n \in rad(B)^2$ is valid. This statement is for $k=1$ one part of the following generalization: for all $k \in \mathbb{N}$, $h \in H$ exists a natural number n such that $rad(B)(\operatorname{ad} h)^n \subseteq rad(B)^{k+1}$ is valid. Within the proof based on induction we assume for natural numbers k, m that $rad(B)(\operatorname{ad} h)^m \subseteq rad(B)^{k+1}$ is valid. Let $b \in rad(B)$, $b' \in rad(B)^k$. The following identity is true:

$$(bb')(\operatorname{ad} h)^{2m} = \sum_{i=0}^{2m} \binom{2m}{i} b(\operatorname{ad} h)^i b'(\operatorname{ad} h)^{2m-i} \in (rad(B))^{k+2}.$$

$rad(B)^{k+1}$ is the linear span of finite many elements bb'. Thus, a natural number l exists such that $rad(B)^{k+1}(\operatorname{ad} h)^l \subseteq rad(B)^{k+2}$ is true. Let $n := m+l$. By using the nilpotency of $rad(B)$ we conclude that B is contained in the Fitting-Null component $B_0(\operatorname{ad} h)$ for every $h \in H$. Hence, we deduce $B \subseteq B_0(\operatorname{ad} H)$. H is a torus, and thus $C_B(H)$ is a Cartan subalgebra based on theorem 14. A Cartan subalgebra is identical to its Fitting-Null component, and we conclude $B_0(\operatorname{ad} H) = C_B(H)$.⋄

Based on the previous lemma we are able to determine the Carter subgroups:

Theorem 15 *(T. Bauer, S. Siciliano) Let A be a finite-dimensional solvable unitary associative F-algebra possessing a separable factor algebra by its nilradical. A subset C of $E(A)$ is a Carter subgroup of $E(A)$ if and only if there exists an algebra complement T of $rad(A)$ in A such that $C = C_{E(A)}(E(T))$ is valid.*

Proof. Let T be an algebra complement of $rad(A)$ in A. By using $E(A) = E(T) \ltimes (1+rad(A))$ we derive for every $k \geq 2$ the condition $\gamma_k(C_{E(A)}(E(T))) \subseteq \gamma_k(C_{1+rad(A)}(E(T)))$. Part (1) of lemma 3 implies the nilpotency of $1+rad(A)$, and hence $C_{E(A)}(E(T))$ is nilpotent, too. Let $g \in N_{E(A)}(C_{E(A)}(E(T)))$ and $B := \langle C_{E(A)}(E(T)) \rangle_F$. By using part (3) of lemma 3 we derive that $\langle E(T) \rangle_F$ is an algebra complement of $rad(B)$ in B. The condition $E(T)^g \subseteq$

$C_{E(A)}(E(T))$ is valid, and thus $(\langle E(T) \rangle_F)^g$ is an algebra complement of $rad(B)$ in B. The theorem of Wedderburn-Malcev implies that $E(T)$ and $E(T)^g$ are conjugated under $1 + rad(B)$. Because of $\langle E(T) \rangle_F \subseteq Z(B)$ we conclude $\langle E(T) \rangle_F = (\langle E(T) \rangle_F)^g$, and hence $E(T)^g = E(T)$ is valid. We obtain
$$[g, E(T)] \subseteq (1 + rad(A)) \cap E(T) = \{1\},$$
and thus $g \in C_{E(A)}(E(T))$ is valid. We conclude that $C_{E(A)}(E(T))$ is a Carter subgroup of $E(A)$.

Now let C be a Carter subgroup of $E(A)$ and $B := \langle C \rangle_F$. By using lemma 4 we conclude that B is Lie nilpotent. Theorem 14 implies that $B = C_B(T_0)$ is valid for an algebra complement T_0 of $rad(B)$ in B. Because of $B = \langle C \rangle_F \subseteq \langle E(B) \rangle_F$ we conclude by part (3) of lemma 3 the condition $T_0 = \langle E(T_0) \rangle_F$. Hence, $E(B) = E((C_B(T_0))) = C_{E(B)}(E(T_0))$ is valid, and by the proof of the first part of this theorem we conclude that this is a Carter subgroup of $E(B)$. Thus, $E(B)$ is nilpotent. The Carter subgroup C is maximal nilpotent in $E(B)$, and we conclude $E(B) = C$. By using part (4) of lemma 3 we derive $rad(B + rad(A)) = rad(A)$. Hence, T_0 is a algebra complement of $rad(B + rad(A))$ in $B + rad(A)$. Again by using the first part of the proof we conclude that $C_{E((B+rad(A)))}(E(T_0))$ is a Carter subgroup of $E((B + rad(A))) = E(T_0) \ltimes (1 + rad(A))$. Because of $C \leq C_{E((B+rad(A)))}(E(T_0))$ we derive $C = C_{E((B+rad(A)))}(E(T_0))$. This implies
$$[C_{E(A)}(E(T_0)), C] \leq C_{E(A)}(E(T_0)) \cap (1 + rad(A)) \leq C,$$
and thus $C_{E(A)}(E(T_0)) \subseteq N_{E(A)}(C) = C$ is valid. By using part (2) of theorem 14 we derive that T_0 is a torus of B. Let T be a maximal torus of A containing T_0. By using part (2) of theorem 14 we conclude that T is an algebra complement of $rad(A)$ in A. Again using the first part of the proof we derive that $C_{E(A)}(E(T))$ is a Carter subgroup of $E(A)$, and thus $C_{E(A)}(E(T)) \leq C_{E(A)}(E(T_0)) \leq C$ is valid. By using the maximal nilpotency of $C_{E(A)}(E(T))$ in C we derive $C = C_{E(A)}(E(T))$, and the proof is finished. ◇

Within infinite solvable groups not all Carter subgroups need to be conjugated. However, within our context we are able to prove the following conjugacy-theorem based on the theorem of Wedderburn-Malcev:

Corollary 2 *(T. Bauer, S. Siciliano) Let A be a finite-dimensional solvable unitary associative K-algebra possessing a separable factor algebra by its nilradical. $E(A)$ possesses exactly one conjugacy class of Carter subgroups.*

Proof. Theorem 15 and the theorem of Wedderburn-Malcev imply the existence of a Carter subgroup of $E(A)$. Let C_1 and C_2 two Carter subgroups of $E(A)$. By using the same theorems we derive the existence of radical complements T_1 and T_2 of $rad(A)$ in A such that $C_i = C_{E(A)}(E(T_i))$ for every $i \in \underline{2}$ is valid. In addition, T_2 and T_1 are conjugated by an element $1 + x \in 1 + rad(A)$: $T_2 = (T_1)^{1+x}$. We conclude $C_2 = C_{E(A)}(E(T_2)) = C_{E(A)}(E((T_1)^{1+x})) = C_{E(A)}(E(T_1)^{1+x}) = C_{E(A)}(E(T_1))^{1+x} = (C_1)^{1+x}$.⋄

Remark 2 *(Bauer, Siciliano) Let $A = K_1 \oplus K_2 \oplus \cdots \oplus K_n$ such that every K_i is a field extension of the field K possessing at least three elements for every $i \in \underline{n}$. A is K-spanned by $E(A)$.*

Proof. Let $a_i \in K_i$ for every $i \in \underline{n}$. The element $(a_1, \cdots, a_n)$ is equal to $\sum_{i=1}^{n}(0, \cdots, 0, a_i, 0, \cdots 0)$. We have to prove that for every $i \in \underline{n}$ the element $(0, \cdots, 0, a_i, 0, \cdots 0)$ is generated by units. If $a_i = 0$, then the zero element is generated by units. Let $0 \neq a_i \neq 1$. The identity $(0, \cdots, 0, a_i, 0, \cdots 0) = (1, \cdots 1, 1 - a_i, 1, \cdots, 1) + (-1, \cdots, -1, a_i, -1, \cdots, -1)$ is valid and both summands are units. Now let $a_i = 1$ and let a an element of K_i different from 0 and 1. In this case it is straightforward to verify the identity $(0, \cdots, 0, 1, 0, \cdots 0) = (1, \cdots 1, 1 - a, 1, \cdots, 1) + (-1, \cdots, -1, a, -1, \cdots, -1)$, and both summands are units. ⋄

Corollary 4 *Let K be a field with at least three elements and A be a finite-dimensional solvable unitary associative K-algebra possessing a separable factor algebra by its nilradical. A is K-generated by $E(A)$.*

Proof. Let T be a radical complement based on the theorem of Wedderburn-Malcev. Then T is semisimple, commutative and unital, and we can apply remark 2 on T. Thus, T is K-generated by $E(T)$. T and $rad(A)$ are decomposing A semidirect, and hence $E(A)$ is composed by $E(T)$ and $1 + rad(A)$ semidirect, too. Now we derive that A is K-generated by $E(A)$. ⋄

Theorem 16 *(T. Bauer, S. Siciliano) Let K be a field possessing at least three elements and A be a finite-dimensional solvable unitary associative K-algebra possessing a separable factor algebra by its nilradical. The Carter subgroups of $E(A)$ are exactly the group of units of the Cartan subalgebras of $A°$. The Cartan subalgebras are exactly the associative subalgebras of A which are K-generated by the Carter subgroups of $E(A)$. Carter subgroups and Cartan subalgebras possess the same class of nilpotency.*

Proof. Let T be an algebra complement of $rad(A)$ in A. By using remark 2 we derive $T = \langle E(T) \rangle_K$, and hence $E((C_A(T))) = C_{E(A)}(E(T))$ is valid. The proof is completed with respect to the theorems 14 and 15 as well as remark 4. The class of nilpotency is identical because of the theorem of Xiankun Du (see theorem 3) and the observation that $C_A(T) = C_{rad(A)}(T) \oplus T$ is valid.⋄

Remark 3 *(Bauer, Siciliano)* The previous result is wrong for a field possessing exactly two elements. For this, we focus on the solvable associative algebra A of upper triangular matrices over a field possessing exactly two elements in dimension 3. The set of diagonal matrices is a Cartan subalgebra, but its group of units is the trivial group. The trivial group is no Carter subgroup because it is not maximal nilpotent: for example the nilpotent subgroup $1 + rad(A)^2$ is different from 1 and $E(A)$).⋄

4.3 Further connections in the case of finite fields

In this section we focus on connections between and the numbers of p-Sylow subgroups, Carter subgroups and p'-Hall subgroups for a finite field of characteristic p.

Proposition 5 *Let K be a field, A a finite-dimensional associative unitary K-algebra, g an unit and T an unital K-subalgebra of A. The identity $E(T)^g = E(T^g)$ is valid.*

Proof. We begin the proof by remarking that for every unital K-subalgebra S of A the statement $E(S) = S \cap E(A)$ is true (see e.g. [75] or [77]). Hence, $E(T)^g = (T \cap E(A))^g = T^g \cap E(A)^g = T^g \cap E(A) = E(T^g)$ is valid.⋄

Theorem 17 *Let K be a finite field of characteristic p, A a finite-dimensional associative unitary solvable K-algebra and T radical complement.*

(i) $1 + rad(A)$ is the p-Sylow subgroup of $E(A)$.

(ii) The p'-Hall subgroups are exactly the conjugates of $E(T)$ under $1 + rad(A)$.

(iii) If K is a splitting field for A, then $E(T) \cong E(K)^{dim_K(A/rad(A))}$ is valid.

(iv) The centralizers of the p'-Hall subgroups are exactly the Carter subgroups.

(v) If T is self-centralizing, then the p'-Hall subgroups are exactly the Carter subgroups.

(vi) If K is a splitting field for A, then the p'-Sylow subgroups of $E(T)$ are a n-fold direct product of cyclic groups, such that $n = dim_K(A/rad(A))$ is valid.

Proof. K is finite, and thus K is perfect and the factor algebra by the nilradical of A separable. By using the theorem of Wedderburn-Malcev a radical complement T of A exists. The algebra A is decomposed by $rad(A)$ and T semidirect. Corollary 1.1.8 in [76] implies that $E(A)$ is the semidirect product of the normal subgroup $1 + rad(A)$ and the subgroup $E(T)$. A is solvable, and thus T is commutative. Hence, there exist field extensions $K_1, \cdots, K_r$ of K such that T is isomorphic to the direct sum $K_1 \times \cdots \times K_r$. Again by using corollary 1.1.8 in [76] we conclude that $E(T) \cong E(K_1) \times \cdots \times E(K_r)$ is valid. The theory of finite fields implies that the order of each extension field K_i ($i \in \underline{r}$) is a power of p. $E(K_i) = K_i \setminus \{0\}$ is valid, and hence $E(K_i)$ is p'-group for all $i \in \underline{r}$. $rad(A)$ is a K-space, and hence parts (i), (iii) and the first part of (ii) are proven.

Let H be a p'-Hall subgroup of $E(A)$. By using a theorem of Philipp Hall an element $g \in E(A)$ exists such that $H = E(T)^g$ is valid. T and T^g are radical complements, and thus by using the theorem of Wedderburn-Malcev an element $r \in rad(A)$ exists such that $T^g = T^{1+r}$ is true. Proposition 5 implies the identity $H = E(T)^g = E(T^g) = E(T^{1+r}) = E(T)^{1+r}$

Parts (iv) and (v) are deductable by part (ii) and by theorem 1 in [4].

Part (v) is a consequence of the facts that the multiplicative group of a finite field is cyclic (see e.g. series I), subgroups of cyclic groups are cyclic, too, and Sylow subgroups are compatible with direct products.⋄

An open topic is to determine the Sylow subgroups of the multiplicative group of a finite field which is a cyclic group of order $p^n - 1$ for a prime number p and a natural number n.

Proposition 6 *Let K be a field, A a finite-dimensional associative unitary solvable K-algebra and T a separable radical complement. The identity $N_{E(A)}(E(T)) = C_{E(A)}(E(T))$ is valid. In addition, if $\mid K \mid \neq 2$ is valid, then the statement $C_{E(A)}(E(T)) = E(C_A(T))$ is true.*

Proof. Let $g \in N_{E(A)}(E(T))$. The algebra A is decomposed by the ideal $rad(A)$ and the subalgebra T. By using corollary 1.1.8 in [76] we conclude that $E(A)$ is the semidirect product of the normal subgroup $1 + rad(A)$ and the subgroup $E(T)$. In particular, there exists elements $r \in rad(A)$ and $t \in E(T)$ such that $g = t(1 + r)$ is valid. The commutativity of T lets us deduce that $E(T) = E(T)^g = E(T)^{t(1+r)} = E(T)^{1+r}$ is true. Hence, the statement $E(T)(1 + r) = (1 + r)E(T)$ is valid. Let $s \in E(T)$. Then an element $x \in E(T)$ exists possessing the property $s(1+r) = (1+r)x$, and we conclude $s + sr = x + rx$. By using $sr, rx \in rad(A)$ and $s, x \in E(T) \leq T$ we deduce $s = x$ and $sr = rs$. Thus, r centralizes every element of $E(T)$, and we have proven $g \in C_{E(A)}(E(T))$.

If $\mid K \mid \neq 2$ is valid, then remark 2 implies the identity $T = \langle E(T) \rangle_K$, and hence $C_{E(A)}(E(T)) = E(C_A(T))$ is true.⋄

Theorem 18 *Let K be a finite field of characteristic p, A a finite-dimensional associative unitary solvable K-algebra and T a radical complement.*

(i) There are exactly $\frac{|E(A)|}{|C_{E(A)}(E(T))|}$ p'-Hall subgroups in $E(A)$. In particular, this number is $\frac{|rad(A)|}{|C_{rad(A)}(E(T))|}$ and is a divisor of the order of the p-Sylow subgroup $1 + rad(A)$ of $E(A)$.

(ii) If T is self-centralizing and $|K| \neq 2$, then there are exactly $|rad(A)|$ p'-Hall subgroups in $E(A)$.
(This number is the maximal possible number of p'-Hall subgroups.)

(iii) The number of p'-Hall subgroups is the same as the number of Carter subgroups of $E(A)$.

Proof. We begin the proof by remarking that the semidirect decomposition of the K-algebra $A = rad(A) \oplus T$ leads (with respect to lemma 3) to a semidirect decomposition of the group of units $E(A) = (1 + rad(A)) \ltimes E(T)$.

ad(i): This is a direct consequence of the preliminary remark, proposition 6 and basic facts of the theory of group operation on sets.

ad(ii): Part (i) and proposition 6 imply the first part of (ii). The second part is a consequence of the commutativity of T.

ad(iii): Carter subgroups are self-normalizing. Thus part (iii) is deductable by part (i), theorem 15 and basic facts of the theory of group operation on sets.◇

4.4 Standard examples

4.4.1 Group algebras

Remark 4 Within section 3.4.1 we have clarified on what terms for a non-abelian group G and for a field K the group algebra KG is solvable: this is valid if and only if $char(K) = p$ is true and G possesses a normal p-Sylow subgroup P. This Sylow subgroup possesses a Hall complement H, and thus KH is a separable commutative radical complement in KG. The nilradical is $KGAug(KP) = Aug(KP)KG$. In the case $p \neq 2$ (see theorem 14) the centralizers of KH and its conjugates under $1 + rad(KG)$ are exactly the Cartan subalgebras of $(KG)^\circ$. The centralizers are defined by the group action of H per conjugation on G: the orbit-sums of the corresponding orbits of H on G is a K-basis of the centralizer of KH in KG. Their group of units are – by using theorem 16 – are exactly the Carter subgroups of $E(KG)$.

If K is finite, then theorems 17 and 18 let us deduce that $1 + rad(KG)$ is the p-Sylow subgroup of $E(KG)$, and the group of units of the radical complements – which are the conjugates of KH under $1 + rad(KG)$ –

are exactly the p'-Hall subgroups of $E(KG)$. Their number is identical to the number of the Carter subgroups and can be calculated by theorem 18 theoretically.◇

A direct consequence of lemma 4 and of remark 4 is:

Corollary 5 *Let K be a field, G a finite group and C a subgroup of G. If KG is solvable, then the following statements are valid:*

(i) If C is nilpotent, then KC is Lie-nilpotent.

(ii) If C is a Carter subgroup of G, then KC is Lie-nilpotent.◇

Whether KC is a Cartan subalgebra for a Carter subgroup C will be clarified later within this section. We focus on the connections between Carter subgroups of G and $E(KG)$ in the case of a solvable associative group algebra KG.

Lemma 5 *(T. Bauer, S. Siciliano) Let G be a finite group and N a normal Sylow subgroup such that G/N is abelian and $C \subseteq G$ is valid. C is a Carter subgroup of G if and only if a complement H of N in G exists such that $C = C_G(H)$ is valid.*

Proof. We begin the proof by remarking that the special properties for the solvability of G imply that the Carter subgroups of G coincide with the so-called Sylow normalizers of G (see e.g. 9.5.10 in [51]).

Let H be a complement of N in G. $G = H \ltimes N$ is valid, and H is an abelian subgroup of G. Let $p_1, p_2, \ldots, p_r$ the distinct prime divisors of $|H|$, and for every $i \in \underline{r}$ let P_i the unique p_i-Sylow subgroup of H. $\{P_1, \ldots, P_r, P_{r+1} = N\}$ is a Sylow basis of G. Let C be the normalizer of this Sylow basis. We obtain

$$C = \bigcap_{i=1}^{r+1} N_G(P_i) = \bigcap_{i=1}^{r} N_G(P_i).$$

Let $i \in \underline{r}$ and $g \in N_G(P_i)$. $[g, P_i] \leq P_i \cap N = \{1\}$, is valid, and hence we conclude $g \in C_G(P_i)$. Thus, $H = P_1 P_2 \ldots P_r$ is true and we deduce

$$C = \bigcap_{i=1}^{r} C_G(P_i) = C_G(H).$$

Hence, the first part of the lemma is proven.

Let C be a Carter subgroup of G. The theorem of Schur-Zassenhaus (see e.g. [51], 9.1.2) implies the existence of a complement H of N in G. Let $\tilde{C} := C_G(H)$. By using the first part of the proof we conclude that $\tilde{C}$ is a Carter subgroup of G. By using a theorem of Roger Carter we deuce that $\tilde{C}^x = C$ is valid for an element $x \in G$. We conclude $C = (C_G(H))^x = C_G(H^x)$, and hence the lemma is proven.◇

Theorem 19 *Let K be a field and G a finite group such that KG is solvable. The following statements are valid:*

(i) Every Carter subgroup C of G is contained in a Carter subgroup $\hat{C}$ of $E(KG)$. In addition, $C = \hat{C} \cap G$ is valid. (T. Bauer, S. Siciliano)

(ii) Every Carter subgroup of $E(KG)$ is an extension of a Carter subgroup of G.

Proof. We use the statements of the preliminary remark 4. If G is abelian, then the theorem is straightforward to prove. Let G be non-abelian. KG is a solvable associative algebra possessing a separable factor algebra by its nilradical.

ad(i): $char(K) = p$ is valid, and G possesses a normal p-Sylow subgroup P and an abelian factor group G/P. Let H be a complement of P in G. By using lemma 5 we obtain that $C := C_G(H)$ is a Carter subgroup of G, and all Carter subgroups are creatable by this process. KH is a separable radical complement of $rad(KG)$ in KG. Let $\tilde{C} = C_{E(KG)}(E(KH))$. By using theorem 15 the subgroup $\tilde{C}$ is a Carter subgroup of $E(KG)$ containing $C = C_G(H)$.
For every subgroup H of G the identity $C_{E(KG)}(E(KH)) \cap G = C_G(H)$ is valid.

ad(ii): Part (ii) is a consequence of part (i) and the conjugacy of Carter subgroups (see corollary 2). ⋄

Within corollary 5 we have proven for a solvable group algebra KG that for every Carter subgroup C of G the subalgebra KC is Lie-nilpotent. But KC is not a Cartan subalgebra which will be proven now:

Proposition 7 *Let K be a field, G a finite group, C a Carter subgroup and U a proper subgroup of G. If KG is solvable, then the following statements are valid:*

(i) KC is a Cartan subalgebra of $(KG)^\circ$ if and only if G is abelian.

(ii) $(KU)^\circ$ is not maximal nilpotent. In particular, every Cartan subalgebra of $(KG)^\circ$ is no (inner) group algebra KU.

Proof. ad(i)+(ii): Let $(KU)^\circ$ be maximal nilpotent. $(KU)^\circ$ is containing the center of KG which is the K-linear span of all conjugacy class sums. Let $g \in G \setminus Z(G)$. Thus, $\overline{g^G}$ is a K-linear sum of elements of U. By comparing coefficients in KG we deduce $g \in U$. It is straightforward to prove that every central element of G is contained in KU. Again by comparing coefficients in KG we conclude $U = G$, and the proposition is proven.⋄

A close connection of Carter subgroups and Cartan subalgebras as proven for solvable associative algebras is not valid for central division algebras. We have proven within series I that the separable maximal subfields are exactly the Cartan subalgebras. But the following statement is valid for Carter subgroups:

Proposition 8 *(T. Bauer, S. Siciliano) Let D be a central division algebra of dimension 4 over a field K of characteristic $\neq 2$. $E(D)$ does not possess an abelian Carter subgroup.*

Proof. We assume that C is an abelian Carter subgroup of $E(D)$. By using a theorem of William Raymond Scott (see theorem 10) the group of units of D is not nilpotent, and thus C possesses an element $y \in E(D) \smallsetminus K$ (Otherwise C would be not only self-normalizing and nilpotent, but also central and thus equal to $E(D)$.). The subalgebra $K[y]$ of D is 2-dimensional and hence a maximal subfield of D. Thus, (maximal subfields are self-centralizing) $C \subseteq C_{E(D)}(y) = C_{E(D)}(K[y]) = E(K[y]) \subseteq N_{E(D)}(C) = C$ is valid. We conclude $C = E(K[y])$ and $N_{E(D)}(K[y]) = E(K[y])$. As the characteristic of K is not 2 the field $K[y]$ is a Galois extension of K (see e.g. the chapter containing topics about quaternion algebras in this work). Thus, an automorphism of order 2 of $K[y]$ exists acting as identity on K. By using the Skolem-Noether theorem (see e.g. 12.6 in [50]) this automorphism is induced by a conjugation based on an element $x \in E(D)$. We derive $x \in N_{E(D)}(K[y]) \smallsetminus E(K[y])$, and this is a contradiction.⋄

4.4.2 The algebras of upper and lower triangular matrices

Let K be a field, $n \in \mathbb{N}$ and $A = \delta_{u,n}$ resp. $A = \delta_{o,n}$. Within both algebras the subalgebra $D(n, K)$ is a self-centralizing radical complement of dimension n. By using theorem 14 the conjugates of $D(n, K)$ by $1 + rad(A)$ are exactly the Cartan subalgebras of $A°$, and theorem 16 implies – in the case of uneven characteristics – that their group of units are exactly the Carter subgroups of $E(A)$. The class of nilpotency of $1 + rad(A)$ is exactly n.

In the case of a finite field of uneven characteristic theorems 18 and 17 let us derive that these Carter subgroups are exactly the p'-Hall subgroups. All Carter subgroups are isomorphic to $E(K)^n$. The quantity of all Carter subgroups is exactly $\mid rad(A) \mid = \mid K \mid^{\frac{n(n-1)}{2}}$. The p-Sylow subgroup is $1 + rad(A)$. Thus, the quantity of all Carter subgroups is exactly the order of the p-Sylow subgroup. The class of nilpotency of $1 + rad(A)$ is determinable by using the theorem of Xiankun Du and theorem 7, and it is exactly n.⋄

4.4.3 Solomon algebras in characteristic zero

Let K be a field of characteristic zero, $n \in \mathbb{N}$ and $A := D_n$. Within the Solomon algebra there exists a self-centralizing radical complement H_n of dimension $p(n)$. Hence – by using theorem 14 – the conjugates of H_n under $1 + rad(A)$ are exactly the Cartan subalgebras of A°, and theorem 16 implies that their group of units are exactly the Carter subgroups of $E(A)$. The class of nilpotency of $1 + rad(A)$ is exactly $n - 1$.$\diamond$

4.4.4 Solomon-Tits algebras

Let K be a field, $n \in \mathbb{N}$ and $A := K\Pi_n$. Within the Solomon-Tits algebra there exists a self-centralizing radical complement V_n of dimension $B(n)$. Hence – by using theorem 14 – the conjugates of V_n under $1 + rad(A)$ are exactly the Cartan subalgebras of A°, and by theorem 16 – in the case of $char(K) \neq 2$ – their group of units are exactly the Carter subgroups of $E(A)$.

In the case of a finite field of uneven characteristic theorems 18 and 17 let us derive that these Carter subgroups are exactly the p'-Hall subgroups. All Carter subgroups are isomorphic to $E(K)^{B(n)}$. The quantity of all Carter subgroups is exactly $\mid rad(A) \mid = \mid K \mid^{\sum_{k=0}^{n}(k!-1)\,S(n,k)}$. The p-Sylow subgroup is $1 + rad(A)$. Thus, the quantity of all Carter subgroups is exactly the order of the p-Sylow subgroup. The class of nilpotency of $1 + rad(A)$ is determinable by using the theorem of Xiankun Du and theorem 7, and it is exactly n.$\diamond$

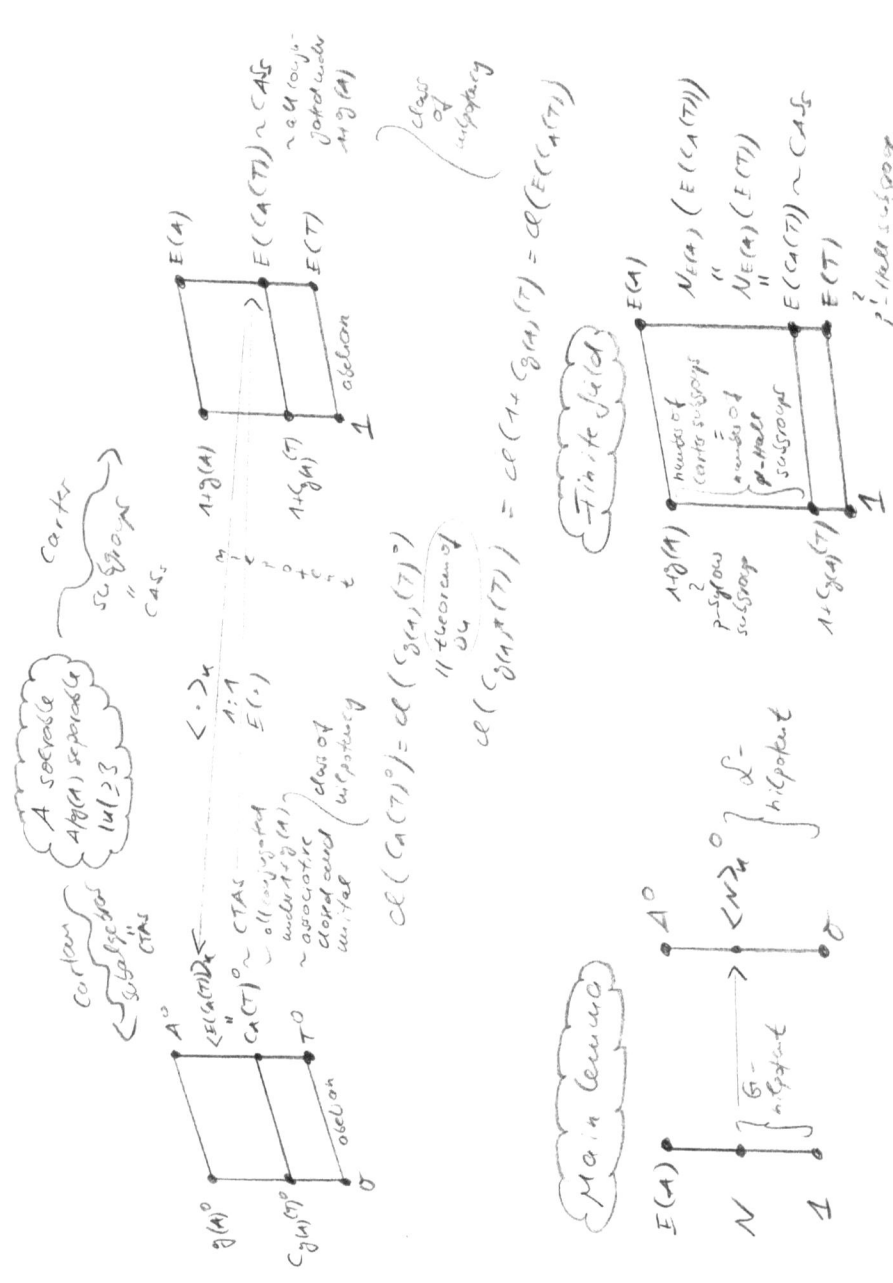

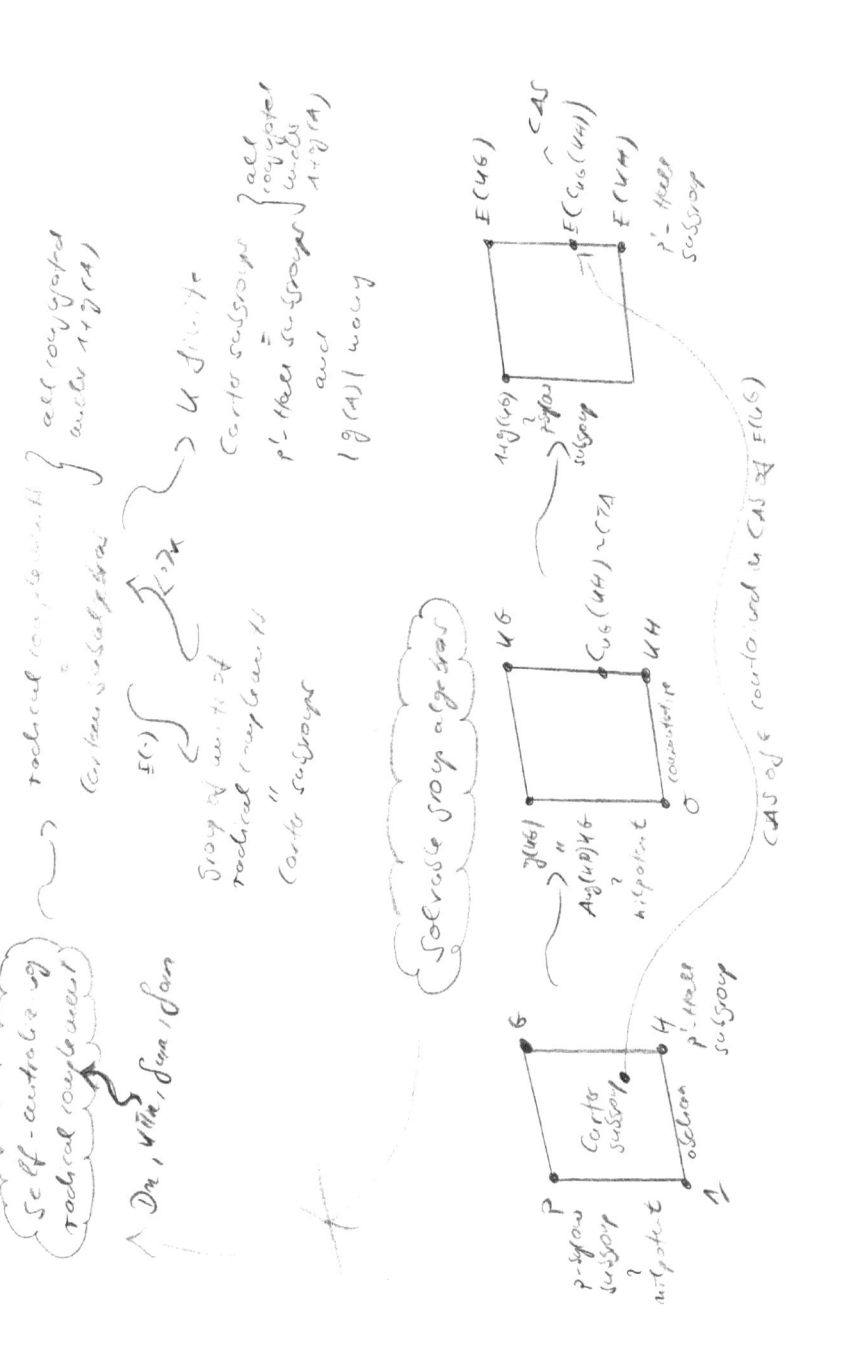

4.5 Open-ended questions and exercises

Open-ended questions 3 *(i) For a solvable group algebra analyze if Cartan subalgebras are isomorphic to group algebras!*

(ii) What are the results of this chapter if the radical factor algebra is not separable?

(iii) Is the extension $\hat{C}$ within theorem 19 unique?

(iv) What are the ascending central chains of the radicals of the Solomon-Tits algebra in arbitrary and the Solomon algebra in characteristic zero?

(v) For a prime number p and a natural number n find the prime factorization of $p^n - 1$.

Excercise 65 *Let K be a finite field of characteristic p and $G := D_{2p}$ or $G := Q_{4p}$. Analyze the following questions:*

(i) What are the Cartan subalgebras of $(KG)^\circ$? What is their dimension?

(ii) What are the p-Sylow subgroups of $E(KG)$? What is their order? Is it possible to calculate or bound their class of nilpotency?

(iii) How many p'-Hall subgroups of $E(KG)$ exist? To which group are all of them isomorphic?

(iv) How many Carter subgroups of $E(KG)$ exist? To which group are all of them isomorphic?

(Tip: series I)

Excercise 66 *Transfer the results of section 4.2 to non-unital solvable associative algebras (Tip: series I, star group, adjunction of an unit.).*

Excercise 67 *Prove that a (semi-)direct decomposition of an associative algebras implies a (semi-)direct decomposition of the star group resp. group of units of the algebra. Is the opposite implication true, too? For group algebras analyze the implication for a (semi-)direct decomposition of the underlying group.*

Excercise 68 *Apply theorems 17 and 18 to the algebras of lower triangular matrices and to the Solomon-Tits algebras for the fields $GF(2^6)$, $GF(3^5)$, $GF(5^4)$, $GF(7^3)$, $GF(11^2)$ and for $n \leq 5$. What do we know about the structure of the Sylow-subgroups of the group of units?*

Excercise 69 *Formulate and prove the following statement in details:*

...The last statement is a consequence of the facts that the multiplicative group of a finite field is cyclic, that subgroups of cyclic groups are cyclic and that Sylow subgroups and direct products are compatible.

For what reason is this statement used within this chapter?

Excercise 70 *Let G be a group, H, U subgroups of G and $g \in G$. The identity $C_G(H)^g = C_G(H^g)$ is valid. Determine $C_U(H)^g$! (Tip: $C_U(H) := C_G(H) \cap U$) For what reason is this statement used within this chapter?*

Excercise 71 *Let A be a finite-dimensional associative unitary K-algebra, T an unital subalgebra of A and g an unit of A. The identity $E(T^g) = E(T)^g$ is valid. How does this identity change for an automorphism α of A instead of the conjugation with g?*

Excercise 72 *Apply the results of this chapter to direct products of algebras!*

Excercise 73 *Apply the results of this chapter to the inverse algebra!*

Excercise 74 *(eAe) Apply the results of this chapter to exercise 60!*

Excercise 75 *(zero-extension) Apply the results of this chapter to exercise 61!*

Excercise 76 *Determine the ascending central chain of the star group for the radical of the algebra of lower triangular matrices. For a finite field analyze the factor groups a long the ascending central chain. To what group is every factor group isomorphic? What about the center itself? If this exercise is to complex, then start by using small values of K and n. What about the descending central chain? What about the series of commutators?*

Excercise 77 *Determine the ascending central chain of the star group for the radical of the algebra of upper triangular matrices. For a finite field analyze the factor groups a long the ascending central chain. To what group is every factor group isomorphic? What about the center itself? If this exercise is to complex, then start by using small values of K and n. What about the descending central chain? What about the series of commutators? Is this exercise connected to exercise 76?*

Excercise 78 *Prove that a group algebra based on a solvable group needs not to be solvable or Lie-solvable. What is the answer for the group of units?*

Excercise 79 *Let p be a prime number different to 2, K a field possessing p elements and $G := D_{2p}$. Analyze (by using the results of series I for Cartan subalgebras) whether the Carter subgroups and p'-Hall are identical. Is it possible to calculate an upper bound for their quantities? Determine both types of subgroups! Is it possible to calculate an upper bound for their class of nilpotency? Is it possible to determine their class of nilpotency?*

Excercise 80 Let p be a prime number different to 2, K a field possessing p elements and $G := D_{4p}$. Analyze (by using the results of series I for Cartan subalgebras) whether the Carter subgroups and p'-Hall are identical. Is it possible to calculate an upper bound for their quantities? Determine both types of subgroups! Is it possible to calculate an upper bound for their class of nilpotency? Is it possible to determine their class of nilpotency?

Excercise 81 Let p be a prime number different to 2, K a field possessing p elements and $G := Q_{4p}$. Analyze (by using the results of series I for Cartan subalgebras) whether the Carter subgroups and p'-Hall are identical. Is it possible to calculate an upper bound for their quantities? Determine both types of subgroups! Is it possible to calculate an upper bound for their class of nilpotency? Is it possible to determine their class of nilpotency?

Excercise 82 Let G be a group and H a subgroup of G. Prove that H is maximal abelian if and only if H is self-centralizing. Does a pendent of this result exists for associative and for Lie algebras?

Chapter 5

The Fitting subgroup of the group of units of a solvable associative algebra

In this chapter we focus on the determination of the Fitting subgroup of the group of units of a finite-dimensional solvable associative algebra possessing a separable factor algebra by its nilradical. We will prove a connection to the nilradical of the associated Lie algebra. The Lie nilradical was determined in series I, and we sum up this result in the first section.

5.1 The pendant within the associated Lie algebra: the nilradical

Within series I we have proven the following result for solvable algebras:

Theorem 20 *Let K be a field, A an associative finite-dimensional unitary solvable K-algebra possessing a separable factor algebra by its nilradical and Z the unique radical complement of the center of A. $rad(A) \oplus Z = rad(A) + Z(A)$ is the nilradical of A°. In particular, the Lie nilradical is an unital associative subalgebra of A and the unique maximal nilpotent subalgebra of A° containing $rad(A)$. The Lie nilradical is of dimension $dim_K(rad(A)) + dim_K(Z)$.*⋄

In the next section we clarify why the Fitting subgroup is the pendant to the Lie nilradical within the group of units.

5.2 The partner within the group of units: the Fitting subgroup

Definition and remark 3 *Let G be a group. By a theorem of Hans Fit-*

ting products of nilpotent normal subgroups are nilpotent. The Fitting subgroup of G is the biggest nilpotent normal subgroup of G (if it is existing) – denoted by $F(G)$. The second Fitting subgroup $F^2(G)$ of G is defined by the identity $F^2(G)/F(G) = F(G/F(G))$. $F^2(G)$ is a normal subgroup of G, and we can determine the Fitting subgroup of $G/F^2(G)$ etc. Thus, we can define the series of Fitting subgroups $(F^n(G))_{n \in \mathbb{N}}$. Each member $F^n(G)$ is a normal subgroup of G (if they are existing) and is called the nth Fitting subgroup. G is solvable if and only if the series of Fitting subgroups reaches G in finite many steps. In this case the number of members of the Fitting series is the Fitting length of G. One interpretation of the Fitting length is to measure the deviation of a solvable group of being nilpotent.$\diamond$

For the determination of the Fitting subgroup lemma 4 is essential and we prove:

Theorem 21 *Let K be a field, A an associative finite-dimensional unitary solvable K-algebra possessing a separable factor algebra by its nilradical and Z the radical complement of the center of A. The Fitting subgroup of $E(A)$ is the group of units of the nilradical of A° which is $(1 + rad(A)) \times E(Z)$. The identity $cl(F(E(A))) = cl(nil(A^\circ)) = cl(rad(A)^\circ)$ is valid. The Fitting subgroup is a maximal nilpotent subgroup of $E(A)$. The second Fitting subgroup is $E(A)$. The K-space generated by the Fitting subgroup is exactly the nilradical of A°.*

Proof. By theorem 20 the subalgebra $rad(A) \oplus Z$ is the nilradical of A°. $rad(A)$ is a nilpotent ideal and Z is central in A. Hence – by using proposition 1.1.8 in [76] – the group of units of the Lie nilradical is a direct product of the nilpotent normal subgroups $1 + rad(A)$ and $E(Z)$. Thus, the group of units of the Lie nilradical of A° is a nilpotent normal subgroup of $E(A)$. Let N be a nilpotent normal subgroup of $E(A)$. $1 + rad(A)$ is a nilpotent normal subgroup, and by a theorem of Hans Fitting the normal subgroup $N(1 + rad(A))$ is nilpotent, too. We use lemma 4 and conclude that the K-linear span $\langle (1+rad(A))N \rangle_K$ is Lie-nilpotent and containing $rad(A)$. By theorem 20 we derive $\langle N \rangle_K \subseteq \langle (1 + rad(A))N \rangle_K \subseteq rad(A) \oplus Z$. In particular, $N \leq E(\langle N \rangle_K) \leq E(\langle (1 + rad(A))N \rangle_K) \leq E(rad(A) \oplus Z)$ is valid. The factor group modulo the Fitting subgroup is abelian (because A is solvable). Thus, $E(A)$ possesses the Fitting length two. The identity concerning the class of nilpotency is a direct consequence of Xiankun Du's theorem 3. The Fitting subgroup contains the derived subgroup. Therefore every subgroup containing the Fitting subgroup is a normal subgroup of $E(A)$. By definition of the Fitting subgroup there is no proper nilpotent subgroup of $E(A)$ containing the Fitting subgroup.$\diamond$

5.3 Standard examples

5.3.1 The algebras of upper and lower triangular matrices

Let K be a field and $n \in \mathbb{N}$. The algebra $\delta_{o,n}$ of upper resp. $\delta_{u,n}$ of lower triangular matrices of $K^{n \times n}$ possesses as nilradical the subalgebra $s\delta_{o,n}$ of strict upper resp. $s\delta_{u,n}$ of strict lower triangular matrices. Its dimension is $\sum_{i=1}^{n-1} i = \frac{1}{2}(n-1)n$. $\delta_{o,n}$ and $\delta_{u,n}$ are central (The center is exactly the one-dimensional subalgebra spanned by the unit element.). By theorem 20 we conclude:

- $nil(\delta_{o,n}{}^\circ) = s\delta_{u,n} \oplus K \cdot 1_{K^{n \times n}}$
- $dim_K(nil(\delta_{o,n}{}^\circ)) = 1 + \frac{1}{2}(n-1)n$
- $nil(\delta_{u,n}{}^\circ) = s\delta_{u,n} \oplus K \cdot 1_{K^{n \times n}}$
- $dim_K(nil(\delta_{u,n}{}^\circ)) = 1 + \frac{1}{2}(n-1)n.\diamond$

By using the previous result and theorem 21 the group of units of the Lie nilradical is exactly the Fitting subgroup. Its isomorphic to the direct product possessing the factors $1 + rad(A)$ and $(K \setminus \{0\}) \cdot 1$. Its class of nilpotency is – based on Du's theorem – exactly the corresponding one of $1_{K^{n \times n}} + s\delta_{o,n}$ resp. $1_{K^{n \times n}} + s\delta_{u,n}$ which is – with respect to theorem 5 – exactly $n.\diamond$

5.3.2 Solomon algebras in characteristic zero

Let K be a field, $char(K) = 0$, $n \in \mathbb{N}$ and D_n the Solomon algebra. Within series I we have determined its nilradical:

- $nil(D_n{}^\circ) = rad(D_n) \oplus Z(D_n)$
- $dim_K(nil(D_n{}^\circ)) = 2^{n-1} - p(n) + 3$, if n is even
- $dim_K(nil(D_n{}^\circ)) = 2^{n-1} - p(n) + 2$, if n is uneven.$\diamond$

The Fitting subgroup can be described by the previous result and based on theorem 21:

- The Fitting subgroup of $E(D_n)$ is the group of units of the nilradical of $(D_n)^\circ$.
- The Fitting subgroup of $E(D_n)$ is exactly $(1 + rad(D_n)) \times E(Z(D_n))$.

Its class of nilpotency is – based on Du's theorem – exactly the corresponding one of $(1 + rad(D_n))$ which is by using theorem 6 exactly $n - 1$. We remark that T. Bauer determines the group $E(Z(D_n))$ and the subalgebra $Z(D_n)$ within his dissertation [4].$\diamond$

5.3.3 Solomon-Tits algebras

Let K be a field, $n \in \mathbb{N}$ and $K\Pi_n$ the Solomon-Tits algebra. Within series I we have determined its nilradical:

- $nil((K\Pi_n)^\circ) = rad(K\Pi_n) \oplus K \cdot 1_{K\Pi_n}$

- $dim_K(nil((K\Pi_n)^\circ)) = 1 + \sum_{k=0}^{n} (k! - 1) \, S(n, k).\diamond$

The Fitting subgroup can be described by the previous result and based on theorem 21:

- The Fitting subgroup of $E(K\Pi_n)$ is the group of units of the nilradical of $(K\Pi_n)^\circ$.

- If K possesses exactly two elements, then the Fitting subgroup of $E(K\Pi_n)$ is exactly $1_{K\Pi_n} + rad(K\Pi_n)$.

- If K possesses more than two elements, then the Fitting subgroup of $E(K\Pi_n)$ is exactly $(1_{K\Pi_n} + rad(K\Pi_n)) \times ((K \setminus \{0_{K\Pi_n}\}) \cdot 1_{K\Pi_n})$.

Its class of nilpotency is – based on Du's theorem – exactly the corresponding one of $rad(K\Pi_n)^\circ$ which is by using theorem 7 exactly $n.\diamond$

5.3.4 Group algebras

Within series I we have determined the nilradical for solvable group algebras:

Theorem 22 *Let G be a finite group and K a field of characteristic $p > 0$ such that KG is solvable. If H is a p'-Hall subgroup and P the normal p-Sylow subgroup of G, then $(Aug(KP)KG) \oplus K(Z(G) \cap H)$ is the nilradical of $(KG)^\circ$. In particular, its dimension is $\mid G \mid - \mid H \mid + \mid Z(G) \cap H \mid$.*

Based on theorem 21 the unit group of $(Aug(KP)KG) \oplus K(Z(G) \cap H)$ is exactly the Fitting subgroup of $E(KG)$. This group is $(1 + Aug(KP)KG) \times E(K(Z(G) \cap H))$. Its class of nilpotency is by the theorem of Xiankun Du 3 exactly the corresponding one of the Lie algebra $(Aug(KP)KG)^\circ$. The determination of this class of nilpotency is not known by the author. An upper bound is given by the associative nilpotency class of $Aug(KP)KG$. This class coincides with the associative nilpotency class of $Aug(KP)$ (because of $Aug(KP)KG = KGAug(KP)$). Kaoru Motose has done calculations for this nilpotency class within several articles (see e.g. [44], [45] and [46]).$\diamond$

Nilradical

A sor'able
A[g(A) separable
∀ i ≥ 3

F. H. of subgroup

Diagram:
- A^0 — $E(w_i c(A^0)) = \mathcal{F}(E(A))$ — $\mathcal{F}(E(A))$ — $F(A)$
- $g(A)^0$ — $1 + g(A)$ — $F(E(A))$
- $\mathcal{Z}(A)^0_{g(A)}$ — — $E(\mathcal{Z}(A))$
- $g(\mathcal{Z}(A))^0$ — T^0 — — $E(T)$
- $(\mathcal{Z}(A \rtimes T))^0$ — $1 + g(\mathcal{Z}(A))$ — $E(\mathcal{Z}(A \rtimes T))$

$E(\cdot)$

$\langle \mathcal{F}(E(A)) \rangle = w_i c(A)$

$\langle \cdot \rangle_u$

$cl(nil(A^0)) = cl(rad(A^0))$
$\parallel$ theorem of
$cl(rad(A)^*) = cl(A \rtimes rad(A)) = cl(\mathcal{F}(E(A)))$

class of
unipotency of
the nilradical

class of unipotency
of the F.H. of
subgroup

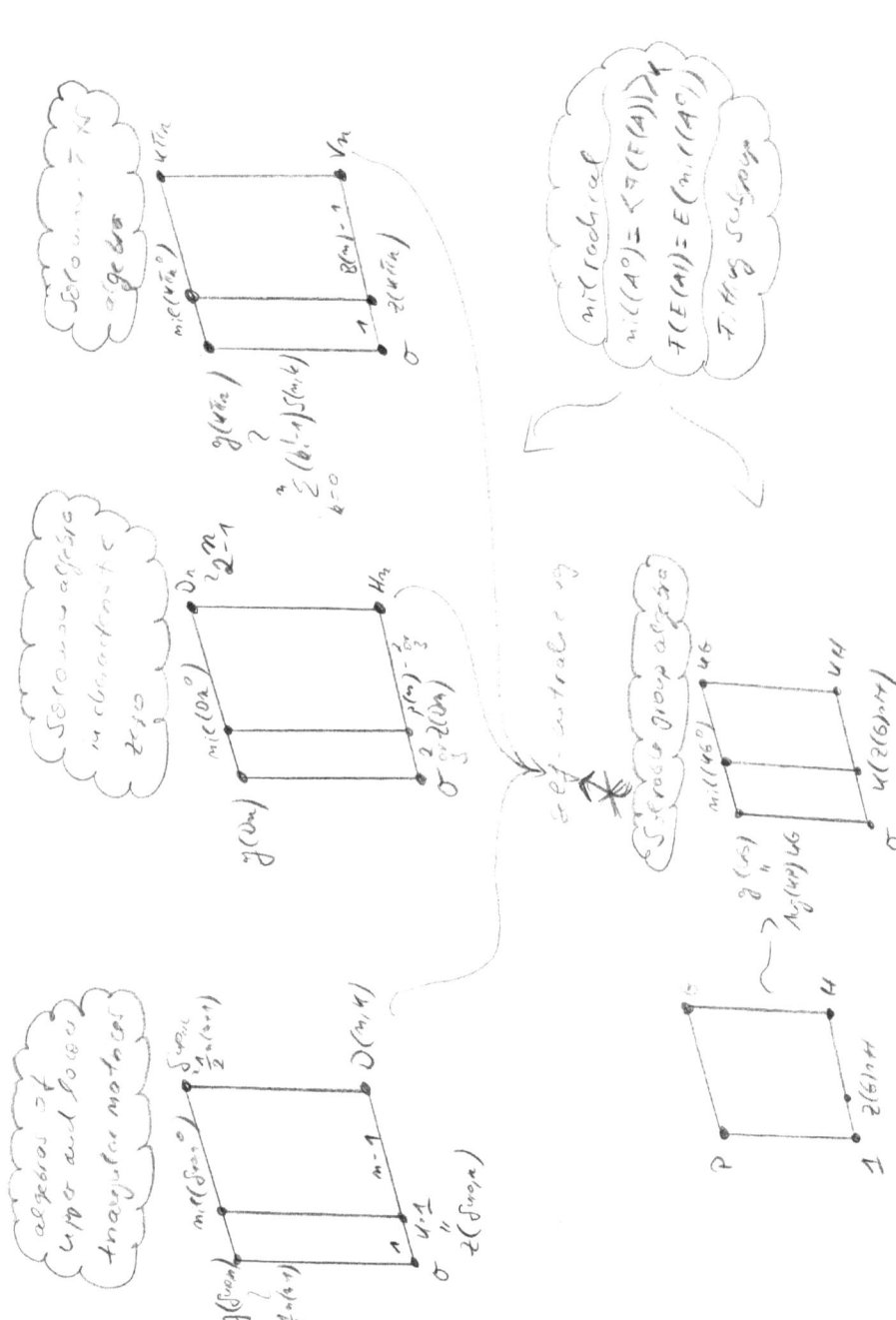

5.4 Open-ended questions and exercises

Open-ended questions 4 *(i) Does a connection exist between the higher Fitting subgroups and suitable Lie ideals?*

(ii) Let G be a group with p-Sylow subgroup P and K a field of characteristic p. What is the Lie nilpotency class of $(Aug(KP)KG)^\circ$ for a solvable group algebra KG?

(iii) Let P be a p-group and K a field of characteristic p. What is the associative and Lie nilpotency class of $Aug(KP)$?

Excercise 83 *Let K be a field. Determine the generalized Jordan decomposition of the matrices $\begin{pmatrix} 1 & 2 & 3 \\ 0 & 1 & 2 \\ 0 & 0 & 1 \end{pmatrix}$ and $\begin{pmatrix} 1 & 2 \\ 1 & 1 \end{pmatrix}$. (Tip: series I)*

Excercise 84 *Let K be a field. Determine the Fitting null component of the linear function based on the matrices $\begin{pmatrix} 1 & 2 & 3 \\ 0 & 1 & 2 \\ 0 & 0 & 1 \end{pmatrix}$ and $\begin{pmatrix} 1 & 2 \\ 1 & 1 \end{pmatrix}$. Determine the Fitting null component of the set of these two linear functions (Tip: series I).*

Excercise 85 *Let K be a field and $n \in \{4, 5\}$. Determine the Lie nilradical and the Fitting subgroup of the algebra of lower triangular matrices of $K^{n \times n}$. What is the dimension of the Lie nilradical? What is the class of nilpotency of the Fitting subgroup? What is the order of the Fitting subgroup for $K = GF(9)$ and for $GF(9^r)$, $r \in \mathbb{N}$?*

Excercise 86 *Let K be a field of characteristic zero and $n \in \{4, 5\}$. Determine the Lie nilradical and the Fitting subgroup of the Solomon algebra D_n. What is the dimension of the Lie nilradical? What is the class of nilpotency of the Fitting subgroup?*

Excercise 87 *Let K be a field of characteristic zero and $n \in \{4, 5\}$. Determine the Lie nilradical and the Fitting subgroup of the Solomon-Tits algebra $K\Pi_n$. What is the dimension of the Lie nilradical? What is the class of nilpotency of the Fitting subgroup? What is the order of the Fitting subgroup for $K = GF(9)$ and for $GF(9^r)$, $r \in \mathbb{N}$?*

Excercise 88 *Apply the main results of this chapter to direct products of algebras!*

Excercise 89 *Apply the main results of this chapter to the inverse algebra!*

Excercise 90 *(eAe) Apply the main results of this chapter to exercise 60!*

Excercise 91 *(zero-extension) Apply the main results of this chapter to exercise 61!*

Excercise 92 *Transfer the main results of this chapter to non-unitary algebras by using the adjunction of an unit and the star group!*

Excercise 93 *Apply the main results of this chapter to exercise 65!*

Excercise 94 *Let p be a prime number different from 2, K a field possessing p elements and $G := D_{2p}$. Determine the Lie nilradical and its dimension as well as the Fitting subgroup and its order.*

Excercise 95 *Let p be a prime number different from 2, K a field possessing p elements and $G := D_{4p}$. Determine the Lie nilradical and its dimension as well as the Fitting subgroup and its order.*

Excercise 96 *Let p be a prime number different from 2, K a field possessing p elements and $G := Q_{4p}$. Determine the Lie nilradical and its dimension as well as the Fitting subgroup and its order.*

Chapter 6

Maximal nilpotency in Lie algebras associated to solvable associative algebras

The aim of this chapter is to determine **all** maximal nilpotent Lie subalgebras in Lie algebras associated to solvable associative algebras with respect to the following questions and topics:

- Is it true that a maximal Lie nilpotent subalgebra is an associative unital subalgebra?

- What is the the inner associative structure of maximal Lie nilpotent subalgebras?

- Is it possible to determine maximal Lie nilpotent subalgebras in a constructive way?

- Is it possible to characterize maximal Lie nilpotent subalgebras by a special property?

- In what way are Cartan subalgebras and the nilradical special among all maximal Lie nilpotent subalgebras?

- Is it possible to bound the number of pairwise non-isomorphic maximal Lie nilpotent subalgebras?

We will answer all of these questions within this chapter. One important instrument for our analysis are single, double and manifold centralizers of subalgebras.

6.1 Associativity

Lemma 6 *Let K be a field, A a finite-dimensional associative (unitary) K-algebra, L a nilpotent Lie-subalgebra of A°, $r \in \mathbb{N}$ and $x, y, l \in A$. The following statements are valid:*

(i) $(xy)\,\mathrm{ad}(l)^r = \sum_{i=0}^{r} \binom{r}{k} x(\mathrm{ad}(l)^k)\, y(\mathrm{ad}(l)^{r-k})$

(ii) $L_0(\mathrm{ad}(l)) := \{x \mid x \in A, \exists n \in \mathbb{N} : x\,\mathrm{ad}(l)^n = 0\}$ *is an associative (unital) subalgebra of A.*

(iii) L *acts nilpotent on the associative algebra-span $\langle L \rangle_{\mathcal{A}}$.*

Proof. ad(i): This is straightforward to verify by using an induction argument and using the derivation property of $\mathrm{ad}(l)$.

ad(ii): Let $a, b \in L_0(\mathrm{ad}(l))$, $c \in K$ and $n, m \in \mathbb{N}$ such that $a(\mathrm{ad}(l)^n) = 0 = b(\mathrm{ad}(l)^m)$ is valid. We define $s := n + m$. It is straightforward to calculate $(a+b)(\mathrm{ad}(l)^s) = 0 = (ca)(\mathrm{ad}(l)^s) = 1(\mathrm{ad}(l)^1)$. By using part (i) we deduce $(ab)(\mathrm{ad}(l)^s) = 0$ (Each summand of the sum in part (i) is zero because one of the factors $x(\mathrm{ad}(l)^k)\, y(\mathrm{ad}(l)^{r-k})$ is always zero.). Hence, $L_0(\mathrm{ad}(l))$ is an associative (unital) subalgebra of A.

ad(iii): This is a consequence of part (ii).◇

Within the following lemma we summarize some facts proven in series I as well as some basic results concerning solvable associative algebras. Let A be an associative K-algebra and $r, s \in A$. A pair $(r; s)$ is called a generalized Jordan decomposition of $r + s$, if r is nilpotent, s is fully-separable and $r \circ s = 0$ is valid (see e.g. series I).

Lemma 7 *Let K be a field, A a finite-dimensional associative (unitary) solvable K-algebra possessing a separable factor algebra by its nilradical, $r, s, a \in A$ and T a(n) (unital) subalgebra of A.*

(i) T *is a finite-dimensional associative (unitary) solvable K-algebra possessing a separable factor algebra by its nilradical.*

(ii) *If $(r; s)$ is a general Jordan decomposition of a, then $(\mathrm{ad}(r); \mathrm{ad}(s))$ is a general Jordan decomposition of $\mathrm{ad}(a)$.*

Proof. ad(i): see [75].

ad(ii): see [79].◇

Now we are in the position to generalize lemma 6 to special solvable algebras:

Lemma 8 *If K is a field, A a finite-dimensional associative solvable (unitary) K-algebra possessing a separable factor algebra by its nilradical and L a nilpotent Lie subalgebra of $A°$, then $\langle L \rangle_A$ is Lie nilpotent, too.*

Proof. By using lemma 6 the Lie subalgebra L acts nilpotent on $\langle L \rangle_A$. The theorem of Friedrich Engel concerning Lie nilpotency lets us deduce that we have to prove for arbitrary elements $x, y \in L$ – assuming that $ad(x), ad(y)$ act nilpotent – also $ad(xy)$ acts nilpotent on $\langle L \rangle_A$. We focus on the unital associative subalgebras $K[x]$ and $K[y]$. Lemma 7 is used to prove that these commutative subalgebras are possessing separable factor algebras by their nilradicals. Hence, x and y possess a generalized Jordan decomposition $(r; s)$ and $(a; b)$. By using the same lemma again $(ad(r); ad(s))$ and $(ad(a); ad(b))$ are generalized Jordan decompositions of $ad(x)$ and $ad(y)$. The restrictions of these ad-functions to $\langle L \rangle_A$ are Jordan decompositions, too. $ad(x)$ and $ad(y)$ act nilpotent on $\langle L \rangle_A$. The uniqueness of the Jordan decomposition lets us deduce that $ad(s)$ and $ad(b)$ are centralizing $\langle L \rangle_A$. We conclude $xy = (r + s)(a + b) = (ra + rb + sa) + sb$. The solvability of A is used to prove that the nilradical of A is exactly the set of nilpotent elements of A (see e.g. series I). Hence, the element $ra + rb + sa$ is nilpotent. s, b are central in $\langle L \rangle_A$, and thus the product sb is central, too. We conclude that $(ra + rb + sa; sb)$ is a generalized Jordan decomposition of xy in $\langle L \rangle_A$ such that sb is central in $\langle L \rangle_A$ is valid. By using lemma 7 the pair $(ad(ra + rb + sa); ad(sb))$ is a generalized Jordan decomposition of $ad(xy)$ (restricted to $\langle L \rangle_A$). sb is central in $\langle L \rangle_A$, and thus $ad(ra + rb + sa)$ is restricted to $\langle L \rangle_A$ identical to $ad(xy)$. We conclude that $ad(xy)$ acts nilpotent on $\langle L \rangle_A$.⋄

By using our previous results we will prove that all maximal Lie nilpotent subalgebras are associative closed within solvable algebras. For Cartan subalgebras and for the Lie nilradical this result was already proven within series I.

Theorem 23 *Let K be a field, A a finite-dimensional associative solvable (unitary) K-algebra possessing a separable factor algebra by its nilradical and L a maximal nilpotent Lie subalgebra of $A°$. L is an associative (unital) subalgebra of A.*

Proof. By using lemma 8 the Lie subalgebra L and the associative subalgebra generated by L are Lie nilpotent. L is maximal Lie nilpotent, and hence L and $\langle L \rangle_A$ are identical. If A is unitary, then let us focus on the Lie subalgebra $L + K1_A$. This subalgebra is Lie nilpotent and is containing L.⋄

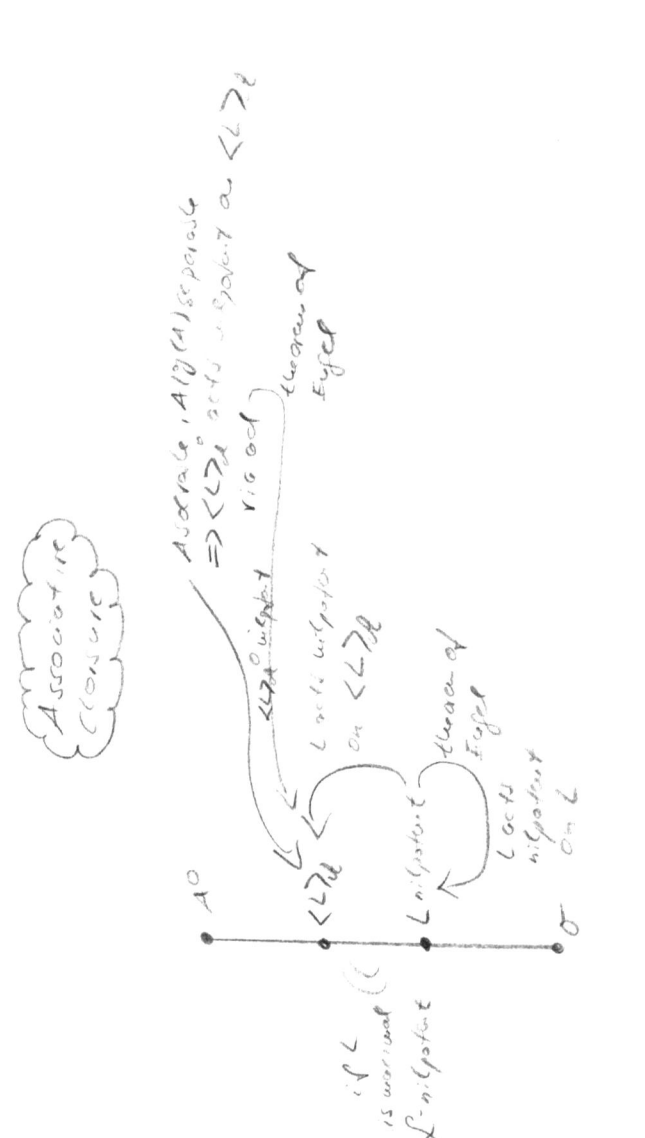

6.2 Manifold centralizers

We begin this section by proving that maximal Lie nilpotent subalgebras possess centralizer and double-centralizer properties:

Lemma 9 *Let K be a field possessing at least three elements, A a finite-dimensional associative unitary solvable K-algebra possessing a separable factor algebra by its nilradical and M a maximal Lie nilpotent subalgebra of A°. A radical complement T of $rad(A)$ in A exists such that the following statements are valid:*

(i) *M possesses a central algebra complement C of $rad(M)$ in M contained in T.*

(ii) *$C_{rad(A)}(C) = rad(M)$*

(iii) *$C_T(rad(M)) = C$*

(iv) *$C_T(C_{rad(A)}(C)) = C$*

(v) *$C_{rad(A)}(C_T(rad(M))) = rad(M)$.*

Proof. ad(i): By using theorem 23 the Lie subalgebra M is an associative K-subalgebra, and M possesses based on lemma 17 exactly one central (and separable) radical complement C. A generalized version of the theorem of Wedderburn-Malcev lets us deduce that C can be enhanced to a radical complement T of A (see e.g. [75]).

ad(ii): By using part (i) the separable subalgebra C is central in M. Hence, the elements of C and $rad(M)$ are pairwise commuting. A is solvable, and thus a theorem in series I (see [75]) lets us deduce that the nilradical of A is exactly the set of nilpotent elements. We conclude $rad(M) \leq C_{rad(A)}(C)$. Let us focus on the K-subspace $C_{rad(A)}(C) \oplus C$ which is containing M. Let $x \in C_{rad(A)}(C)$ and $c \in C$. Then $xc \in rad(A)$ is valid, and by using the commutativity of C it is straightforward to prove that xc centralizes C. Thus, $xc \in C_{rad(A)}(C)$ is valid, and by using a similar argument $cx \in C_{rad(A)}(C)$ is true. Hence, $C_{rad(A)}(C) \oplus C$ is an associative subalgebra and $C_{rad(A)}(C)$ is a nilpotent ideal of $C_{rad(A)}(C) \oplus C$. C is a central radical complement for the nilradical $C_{rad(A)}(C)$. Hence, $C_{rad(A)}(C) \oplus C$ is Lie nilpotent, an by using the maximality of M we deduce $C_{rad(A)}(C) \oplus C = M = C \oplus rad(M)$. Thus, $rad(M) = C_{rad(A)}(C)$ is true, and part (ii) is proven.

ad(iii): By using part (i) we deduce that the subalgebra C is central in M, and thus $C \leq C_T(rad(M))$ is valid. Dual to the proof of part (ii) we focus on the K-subspace $rad(M) \oplus C_T(rad(M))$. This K-space is – by using part (i) – identical to the K-subspace $C_{rad(A)}(C) \oplus C_T(rad(M))$. Let $y \in C_{rad(A)}(C)$

and $x \in C_T(rad(M))$. For all $c \in C$ the rule $xy \circ c = (x \circ c)y + x(y \circ c)$ is valid. Both summands are zero because of the commutativity of T and the definition of $C_{rad(A)}(C)$. Thus, $xy \in C_{rad(A)}(C)$ is true, and by a similar argument we derive $yx \in C_{rad(A)}(C)$. We have proven that $C_{rad(A)}(C) \oplus C_T(rad(M))$ is an associative subalgebra and $C_{rad(A)}(C)$ is a nilpotent ideal of this subalgebra. $C_T(rad(M))$ is a central radical complement of $rad(M)$, and thus the subalgebra $C_{rad(A)}(C) \oplus C_T(rad(M))$ is Lie nilpotent and containing M. By using the maximality of M we deduce part (iii).

ad(iv)+(v): These properties are a direct consequence of parts (ii) and (iii).⋄

The parts (iv) and (v) of the previous lemma can be used to develop an idea how to construct all maximal Lie nilpotent subalgebras. The following definition is related to this idea:

Definition 1 Let K be a field, A a finite-dimensional associative K-algebra possessing a separable factor algebra by its nilradical, T a radical complement of $rad(A)$ in A, C a subalgebra of T and I a subalgebra of $rad(A)$.

We define sequences of subalgebras starting with C resp. I. Let $\mathcal{J}_1(I) := I$ and for $i \geq 2$ we define recursively

$$\mathcal{J}_i(I) := \begin{cases} C_T(\mathcal{J}_{i-1}(I)) &, i \text{ even} \\ C_{rad(A)}(\mathcal{J}_{i-1}(I)) &, i \text{ uneven} \end{cases}.$$

In addition, we define $\mathcal{T}_1(C) := C$ and for all $i \geq 2$ recursively

$$\mathcal{T}_i(C) := \begin{cases} C_{rad(A)}(\mathcal{T}_{i-1}(C)) &, i \text{ even} \\ C_T(\mathcal{T}_{i-1}(C)) &, i \text{ uneven} \end{cases}.$$

The first sequence starts with C, calculates the centralizer Z of C in $rad(A)$, afterwards based on Z the centralizer Y in T and so on.

The second sequence does not start within the radical complement, but within the nilradical with the subalgebra I, calculates the centralizer Z of I in T, afterwards based on Z the centralizer Y in $rad(A)$ and so on.

Both sequences are switching between the radical and the radical complement based on centralizers. In particular, each member of both sequences is an associative subalgebra. At the end of this section a graphic illustrates the sequences for the reader. We will focus on both sequences for solvable associative algebras and analyze their relationship to maximal nilpotent Lie subalgebras. For this, the most interesting fact is that these sequences will be constant after finite many steps. The resulting subalgebras are related to maximal nilpotent Lie subalgebras.⋄

We begin our analysis by proving some basic properties of the sequences defined in definition 1. Except part (xi) - which is a consequence of the previous lemma – all statements are straightforward to prove.

Proposition 9 *Let K be a field, A a finite-dimensional associative K-algebra possessing a separable factor algebra by its nilradical, T a radical complement of $rad(A)$ in A, C, D subalgebras of T and N, M subalgebras of $rad(A)$. The following statements are valid:*

(i) $C \leq C_T(C_{rad(A)}(C))$

(ii) $N \leq C_{rad(A)}(C_T(N))$

(iii) *The subsequences $(\mathfrak{T}_n(C))_{n \in 2\mathbb{N}}$ and $(\mathfrak{T}_n(C))_{n \in 2\mathbb{N}+1}$ are monotonically increasing.*

(iv) *The subsequences $(\mathfrak{J}_n(N))_{n \in 2\mathbb{N}}$ and $(\mathfrak{J}_n(N))_{n \in 2\mathbb{N}+1}$ are monotonically increasing.*

(v) *The subsequences $(\mathfrak{T}_n(C))_{n \in 2\mathbb{N}}$ and $(\mathfrak{T}_n(C))_{n \in 2\mathbb{N}+1}$ are constant for almost all n.*

(vi) *The subsequences $(\mathfrak{J}_n(N))_{n \in 2\mathbb{N}}$ and $(\mathfrak{J}_n(N))_{n \in 2\mathbb{N}+1}$ are constant for almost all n.*

(vii) *If $C \leq D$ is valid, then we derive $C_{rad(A)}(D) \leq C_{rad(A)}(C)$.*

(viii) *If $N \leq M$ is valid, then we derive $C_T(M) \leq C_T(N)$.*

(ix) *If $C \leq D$ is valid, then $\mathfrak{T}_n(C) \leq \mathfrak{T}_n(D)$ is true for all uneven $n \in \mathbb{N}$ and $\mathfrak{T}_n(D) \leq \mathfrak{T}_n(C)$ for all even $n \in \mathbb{N}$.*

(x) *If $N \leq M$ is valid, then $\mathfrak{J}_n(N) \leq \mathfrak{J}_n(M)$ is true for all uneven $n \in \mathbb{N}$ and $\mathfrak{J}_n(M) \leq \mathfrak{J}_n(N)$ for all even $n \in \mathbb{N}$.*

(xi) *Let A be solvable and $\mid K \mid \geq 3$. If $C \oplus N$ is maximal Lie nilpotent, then all four subsequences $(\mathfrak{T}_n(C))_{n \in 2\mathbb{N}}, (\mathfrak{T}_n(C))_{n \in 2\mathbb{N}+1}, (\mathfrak{J}_n(N))_{n \in 2\mathbb{N}}$ and $(\mathfrak{J}_n(N))_{n \in 2\mathbb{N}+1}$ are constant.*

(xii) *If for one $n \in \mathbb{N}$ the identity $\mathfrak{T}_n(C) = \mathfrak{T}_{n+2}(C)$ is valid, then the subsequences $(\mathfrak{T}_n(C))_{n \in 2\mathbb{N}}$ and $(\mathfrak{T}_n(C))_{n \in 2\mathbb{N}+1}$ are constant from the nth resp. $(n+1)$th term.*

(xiii) *If for one $n \in \mathbb{N}$ the identity $\mathfrak{J}_n(N) = \mathfrak{J}_{n+2}(N)$ is valid, the subsequences $(\mathfrak{J}_n(N))_{n \in 2\mathbb{N}}$ and $(\mathfrak{J}_n(N))_{n \in 2\mathbb{N}+1}$ are constant from the nth resp. $(n+1)$th term.*

Proof. ad(i): Let $c \in C$. Hence, $c \in T$ is valid. We have to prove that c commutes with all elements of $C_{rad(A)}(C)$. But this statement is true by definition of $C_{rad(A)}(C)$.

ad(ii): This part is proven as done within part (i).

ad(iii)-(vi): These parts are a consequence of part (i) and (ii) as well as using the finite dimension of A.

ad(vii): All element centralizing D are centralizing $C \leq D$, too.

ad(viii): This part is proven as done within part (vii).

ad(ix)+(x): These parts are a consequence of part (vii) and (viii).

ad(xi): This part is included in lemma 9.

ad(xii)+(xiii): These parts are straightforward to prove.⋄

The next lemma demonstrates us that maximal Lie nilpotent subalgebras are connected to the limits of the subsequences defined and described within proposition 9.

Lemma 10 *Let K be a field possessing at least three elements, A a finite-dimensional associative solvable K-algebra possessing a separable factor algebra by its nilradical, T an algebra complement of $rad(A)$ in A, C a subalgebra of T and N a subalgebra of $rad(A)$. The following statements are valid:*

(i) If $C = C_T(C_{rad(A)}(C))$ is valid, then $C \oplus C_{rad(A)}(C)$ is a maximal Lie nilpotent subalgebra.

(ii) If $N = C_{rad(A)}(C_T(N))$ is valid, then $C_T(N) \oplus N$ is a maximal nilpotent Lie subalgebra.

Proof. ad(i): $C \oplus C_{rad(A)}(C)$ is an associative subalgebra possessing the nilradical $C_{rad(A)}(C)$, and C is a radical complement (see lemma 9). Both subalgebras are centralizing each other, and thus $C \oplus C_{rad(A)}(C)$ is Lie nilpotent. Let M be a maximal nilpotent Lie subalgebra of A° containing $C \oplus C_{rad(A)}(C)$. By using theorem 23 and lemma 6 we conclude that M is an associative subalgebra possessing a central radical complement M_C: M_C is exactly the set of fully-separable elements in M. We derive $M = rad(M) \oplus M_C$ and $C \oplus C_{rad(A)}(C) \leq M$. A is solvable, and thus $C_{rad(A)}(C) \leq rad(M)$ is valid. M_C is central in M, and we conclude $C \leq M_C$ using the enhance conjugacy theorem of Wedderburn-Malcev (see

[75]). Lemma 9 results in $C_{rad(A)}(M_C) = rad(M)$ and $C_T(rad(M)) = M_C$. Thus, $M_C = C_T(rad(M)) \leq C_T(C_{rad(A)}(C))$ is valid. By using the assumptions of this theorem we conduct $M_C \leq C$ which results in $M_C = C$. Hence, $rad(M) = C_{rad(A)}(M_C) = C_{rad(A)}(C)$ is valid, and part (i) is proven.

ad(ii): Part (ii) will be derived from part (i). We focus on $C_T(N)$. By our assumption $C_T(C_{rad(A)}(C_T(N))) = C_T(N)$ is valid. Part (i) lets us conclude that $C_T(N) \oplus C_{rad(A)}(C_T(N))$ is maximal Lie nilpotent. The term on the right hand side is exactly N (by our assumption).⋄

The following theorem lets us deduce that maximal Lie nilpotent subalgebras can be characterized by special centralizer properties. We know already that maximal Lie nilpotent subalgebras are associative subalgebras containing exactly one central radical complement. By using lemma 10 and 9 we conclude:

Theorem 24 *Let K be a field possessing at least three elements, A a finite-dimensional associative solvable unitary K-algebra possessing a separable factor algebra by its nilradical, $M = rad(M) \oplus M_C$ a Lie nilpotent subalgebra possessing a central radical complement M_C and T a radical complement of A containing M_C. The following statements are equivalent:*

(i) M is maximal Lie nilpotent.

(ii) The identities $rad(M) = C_{rad(A)}(M_C)$ and $M_C = C_T(rad(M))$ are valid.

Add-on: If (i) or (ii) is valid, then the statements $C_T(C_{rad(A)}(M_C)) = M_C$ and $C_{rad(A)}(C_T(rad(M))) = rad(M)$ are true.⋄

Within the exercises we highlight an example such that the add-on stated in theorem 24 is not equivalent to the maximal Lie nilpotency. Based on proposition 9 we want to analyze whether the defined sequences – based on special subalgebras – of iterative centralizers are constant after finite many steps. Surprisingly, these sequences are constant latest after the second step for every subalgebra they are based on. By using this result and lemma 10 we are able to construct all maximal Lie nilpotent subalgebras.

Lemma 11 *Let K be a field, A a finite-dimensional associative K-algebra possessing a separable factor algebra by its nilradical, T an algebra complement of $rad(A)$ in A, C a subalgebra of T and N a subalgebra of $rad(A)$. The following statements are valid:*

(i) $C_T(C_{rad(A)}(C_T(C_{rad(A)}(C)))) = C_T(C_{rad(A)}(C))$

(ii) $C_{rad(A)}(C_T(C_{rad(A)}(C_T(N)))) = C_{rad(A)}(C_T(N))$

Proof. ad(i): By using proposition 9, part (i) twice we deduce

$$C_T(C_{rad(A)}(C)) \geq C$$

and thus

$$C_T(C_{rad(A)}(C_T(C_{rad(A)}(C)))) \geq C_T(C_{rad(A)}(C)).$$

The same proposition, part (ii) applied to $N := C_{rad(A)}(C)$ lets us deduce

$$C_{rad(A)}(C_T(N)) \geq N$$

, and this is equivalent to

$$C_{rad(A)}(C_T(C_{rad(A)}(C))) \geq C_{rad(A)}(C).$$

Using $C_T(\cdot)$ on both side of this identity we derive the opposite inclusion

$$C_T(C_{rad(A)}(C_T(C_{rad(A)}(C)))) \leq C_T(C_{rad(A)}(C)).$$

ad(ii): The proof of part (ii) is to be done as an exercise by the reader. Its similar to the one done in part (i). ⋄

Based on the lemmata 9, 10 and 11 we are able to construct all maximal Lie nilpotent subalgebras such that their fully-separable part is contained in a fixed radical complement of the underlying algebra:

Theorem 25 *Let K be a field possessing at least three elements, A a finite-dimensional associative solvable K-algebra possessing a separable factor algebra by its nilradical, T an algebra complement of $rad(A)$ in A, C a subalgebra of T and N a subalgebra of $rad(A)$. The following statements are valid:*

(i) $C_T(C_{rad(A)}(C)) \oplus C_{rad(A)}(C_T(C_{rad(A)}(C)))$ is maximal Lie nilpotent.

(ii) $C_{rad(A)}(C_T(N)) \oplus C_T(C_{rad(A)}(C_T(N)))$ is maximal Lie nilpotent.

Every maximal Lie nilpotent subalgebra such that their fully-separable part is contained in T are constructible by this procedure.⋄

By using this theorem we have to execute the double-centralizing on every unital subalgebra of a fixed radical complement T or on every subalgebra of the nilradical $rad(A)$ once. The resulting subalgebras can be used based on lemma 10 to construct maximal Lie nilpotent subalgebras. All maximal Lie nilpotent subalgebras possessing a fully-separable part contained in a fixed radical complement are constructible by this procedure which is the statement of lemma 9. To focus on the unital subalgebras of T is sufficient: If C is an arbitrary non-unital subalgebra of T, then $C_T(C_{rad(A)}(C)) = C_T(C_{rad(A)}(C \oplus K \cdot 1_A))$ is valid. $C \oplus K \cdot 1_A$ is an unital subalgebra. Both procedures result in the same maximal Lie nilpotent subalgebras based on theorem 24: based on $N \leq rad(A)$ we produce the maximal Lie nilpotent subalgebra $C_{rad(A)}(C_T(N)) \oplus C_T(C_{rad(A)}(C_T(N)))$. If we use $C := C_T(N) \leq T$, then we create the maximal Lie nilpotent subalgebra $C_T(C_{rad(A)}(C_T(N))) \oplus C_{rad(A)}(C_T(C_{rad(A)}(C_T(N))))$. Both subalgebras are identical because $C_{rad(A)}(C_T(C_{rad(A)}(C_T(N)))) = C_{rad(A)}(C_T(N))$ is valid.

The following attractor resp. repeller properties are valid for the double-centralizing:

Corollary 6 *Let K be a field possessing at least three elements, A a finite-dimensional associative solvable K-algebra possessing a separable factor algebra by its nilradical, T an algebra complement of $rad(A)$ in A, C, D subalgebras of T and N, M subalgebras of $rad(A)$. The following statements are valid:*

(i) $C \leq D \leq C_T(C_{rad(A)}(C))$ results in $C_T(C_{rad(A)}(C)) = C_T(C_{rad(A)}(D))$.

(ii) $N \leq M \leq C_{rad(A)}(C_T(N))$ results in $C_{rad(A)}(C_T(N)) = C_{rad(A)}(C_T(M))$.

(iii) $C_T(C_{rad(A)}(C)) = C_T(C_{rad(A)}(D))$ results in $D \leq C_T(C_{rad(A)}(C))$.

(iv) $C_{rad(A)}(C_T(N)) = C_{rad(A)}(C_T(M))$ results in $M \leq C_{rad(A)}(C_T(N))$.

(v) $C_T(C_{rad(A)}(C)) = C_T(C_{rad(A)}(D))$ results in $\langle C \cup D \rangle_A \leq C_T(C_{rad(A)}(C))$.

(vi) $C_{rad(A)}(C_T(N)) = C_{rad(A)}(C_T(M))$ results in $\langle N \cup M \rangle_A \leq C_{rad(A)}(C_T(N))$.

(vii) $C_T(C_{rad(A)}(C)) < D$ results in $C_T(C_{rad(A)}(C)) < C_T(C_{rad(A)}(D))$.

(viii) $C_{rad(A)}(C_T(N)) < M$ results in $C_{rad(A)}(C_T(N)) < C_{rad(A)}(C_T(M))$.

(ix) If D is no subset of $C_T(C_{rad(A)}(C))$, then $C_T(C_{rad(A)}(D))$ is no subset of $C_T(C_{rad(A)}(C))$, too.

(x) If N is no subset of $C_{rad(A)}(C_T(N))$, then $C_{rad(A)}(C_T(M))$ is no subset of $C_{rad(A)}(C_T(N))$, too.

Proof. ad(i): By using proposition 9 and $C \leq D$ the statement

$$C_T(C_{rad(A)}(C)) \leq C_T(C_{rad(A)}(D)).$$

is valid. By an analogue argumentation based on $D \leq C_T(C_{rad(A)}(C))$ we conclude

$$C_T(C_{rad(A)}(D)) \leq C_T(C_{rad(A)}(C_T(C_{rad(A)}(C)))).$$

Lemma 11 results in

$$C_T(C_{rad(A)}(C_T(C_{rad(A)}(C)))) = C_T(C_{rad(A)}(C)),$$

and part (i) is proven.

ad(ii): This statement can be proven by an analogue argumentation as used in part (i) and may be done as an exercise by the reader.

ad(iii): This part is a direct consequence of proposition 9:
$$D \leq C_T(C_{rad(A)}(D))$$
is valid.

ad(iv): This statement can be proven by an analogue argumentation as used in part (iii) and may be done as an exercise by the reader.

ad(v): By using proposition 9 we conclude that C and also D are subalgebras of
$$C_T(C_{rad(A)}(C)) = C_T(C_{rad(A)}(D))$$
, and part (v) is proven.

ad(vi): This statement can be proven by an analogue argumentation as used in part (v) and may be done as an exercise by the reader.

ad(vii): Proposition 9 results in
$$D \leq C_T(C_{rad(A)}(D)),$$
and thus part (vii) is proven.

ad(viii): This statement can be proven by an analogue argumentation as used in part (vii) and may be done as an exercise by the reader.

ad(ix): Proposition 9 results in
$$D \leq C_T(C_{rad(A)}(D)),$$
and part (viii) is proven.⋄

Definition 2 Let K be a field possessing at least three elements, A a finite-dimensional associative unitary solvable K-algebra possessing a separable factor algebra by its nilradical, T a radical complement, $C = C_T C_{rad(A)}(C)$ an unital subalgebra of T and $J = C_{rad(A)} C_T(J)$ a subalgebra of $rad(A)$. The attraction section of C is the set of all subalgebras D of T such that $C_T C_{rad(A)}(C) = C_T C_{rad(A)}(D)$ is valid. The attraction section of J is the set of all subalgebras D of $rad(A)$ such that $C_{rad(A)} C_T(J) = C_{rad(A)} C_T(D)$ is valid.⋄

Remark 5 Let K be a field possessing at least three elements, A a finite-dimensional associative solvable K-algebra possessing a separable factor algebra by its nilradical, T a radical complement of $rad(A)$ in A, C, D subalgebras of T and N, M subalgebras of $rad(A)$. Corollary 6 and lemma 11 lets us construct different sequences of maximal Lie nilpotent subalgebras. In this remark we focus on sequences within the radical complement T. The reader may define analogue sequences within the exercises based on the nilradical. By using lemma 11 – beginning with C – the subalgebra $C_T(C_{rad(A)}(C))$ is suitable to construct a maximal Lie nilpotent subalgebra:

$$C_{rad(A)}(C_T(C_{rad(A)}(C))) \oplus C_T(C_{rad(A)}(C)).$$

We define sequences of subalgebras $(C_n)_{n \in \mathbb{N}}$ such that

$$C_T(C_{rad(A)}(C_n)) = C_n$$

is valid, and thus

$$C_n \oplus C_{rad(A)}(C_T(C_n))$$

are maximal Lie nilpotent subalgebras for all $n \in \mathbb{N}$.

Sequence 1:

We start by using $C \leq Z(A) \cap T$. Thus, $C_{rad(A)}(C) = rad(A)$ is valid, and hence $C_1 := C_T(C_{rad(A)}(C)) = C_T(rad(A)) = Z(A) \cap T$ is true because of the solvability of A (T is commutative) and T being a radical complement. The maximal Lie nilpotent subalgebra associated to this double-centralizer is exactly the nilradical (see theorem 20). Choose a subalgebra D_1 containing $Z(A) \cap T$ proper and minimal with this property, and construct $C_2 := C_T(C_{rad(A)}(D_1))$. C_2 is containing C_1 proper (see corollary 6), and based on C_2 we are able to define another maximal Lie nilpotent subalgebra. Now choose D_2 containing C_2 proper and minimal with this property (if this is possible) and perform the same steps as defined previously for D_1. Because of the finite dimension of T we are reaching T after finite many steps with this construction. The radical complement is associated to a Cartan subalgebra as describe in theorem 14. From now on the sequence is constant.

Sequence 2:

We start by using the radical complement $C_1 := T$. T is – as mentioned within the construction of sequence 1 – associated to a Cartan subalgebra. Choose a subalgebra M_1 as small as possible in T with the property $C_T(C_{rad(A)}(M_1)) = T$. Choose D_1 as a subalgebra of M_1 maximal with the property $C_2 := C_T(C_{rad(A)}(D_1)) \neq T$. This construction is repeated by using C_2 instead of T. By this procedure the resulting subalgebras C_n are getting smaller within each step. Hence, by using the finite dimensionality the sequence of the subalgebras C_n is reaching $Z(A) \cap T$ after finite many steps. As described in sequence 1 the procedure is associated to the nilradical from now on and is constant.

Sequence 3:

Sequence 3 is a mixture of sequence 1 and 2. Choose a subalgebra C which is neither associated to a Cartan subalgebra nor to the nilradical (if such a subalgebra is existing). Starting by C we use the procedure in sequence 1 reaching the radical complement T after finite many steps which is associated to a Cartan subalgebra. From C down to $Z(A) \cap T$ we define another sequence of subalgebras in T analogue to the procedure of sequence 2. This sequence is constant of finite many steps, and the final subalgebra in T is associated to the nilradical.

Cartan subalgebras and the nilradical are in view of these sequence extreme within all maximal Lie nilpotent subalgebras. For the nilradical its part in T is extremely small, but for the Cartan subalgebra extremely big. The reader may define analogue sequencers within the nilradical. For these the situation is dual: for the nilradical the nilpotent part is extremely big, but for the Cartan subalgebras extremely small. The following corollary is related to this phenomena.⋄

Corollary 7 *Let K be a field possessing at least three elements, A a finite-dimensional associative solvable unitary K-algebra possessing a separable factor algebra by its nilradical, T a radical complement of $rad(A)$ in A, C a subalgebra of T and N a subalgebra of $rad(A)$. The following statements are valid:*

(i) $C_T(C_{rad(A)}(C)) = Z(A) \cap T$ is valid if and only if C is a subalgebra of $Z(A) \cap T$.
(attraction section of the nilradical in T; see theorem 25)

(ii) $C_{rad(A)}(C_T(N)) = C_{rad(A)}(T)$ is valid if and only if N is a subalgebra of $C_{rad(A)}(T)$.
(attraction section of the Cartan subalgebras in $rad(A)$; see theorem 25)

(iii) $C_{rad(A)}(C_T(N)) = rad(A)$ is valid if and only if $C_T(N)$ is central in A. (attraction section of the nilradical in $rad(A)$; see theorem 25)

(iv) $C_T(C_{rad(A)}(C)) = T$ is valid if and only if $C_{rad(A)}(C) = C_{rad(A)}(T)$ is valid. (attraction section of the Cartan subalgebras in T; see theorem 25)

Proof. ad(i): Let C be central in A. $C_{rad(A)}(C) = rad(A)$ is valid, and we derive $C_T(C_{rad(A)}(C)) = C_T(rad(A)) = Z(A) \cap T$ because T is a commutative radical complement. Let $C_T(C_{rad(A)}(C)) = Z(A) \cap T$ be valid. C is contained in $C_T(C_{rad(A)}(C))$, and thus part (i) is proven (the add-on is proven by using theorem 20).

ad(ii): By using theorem 14 the subalgebra $T \oplus C_{rad(A)}(T)$ is a Cartan subalgebra of $A°$. In particular, this subalgebra is maximal Lie nilpotent. By using lemma 9 the statement $C_{rad(A)}(T) = C_{rad(A)}(C_T(C_{rad(A)}(T)))$ is valid. Corollary 6 lets us deduce that for every subalgebra N of $C_{rad(A)}(T)$ the property to be proven is valid. If this property is valid for a subalgebra N, then $C_{rad(A)}(C_T(N)) = C_{rad(A)}(T)$ is true. N is contained in this subalgebra, and we conclude $C_T(N) \geq C_T(C_{rad(A)}(T)) = T$. This results in $C_T(N) = T$. Hence, $C_{rad(A)}(C_T(N)) = C_{rad(A)}(T)$ is valid. N is contained in $C_{rad(A)}(C_T(N))$, and by using theorem 14 this part is proven.

ad(iii): $C_{rad(A)}(C_T(N)) = rad(A)$ is valid if and only if $rad(A) \circ C_T(N) = 0$ is true. This statement is equivalent to the statement that for $t \in T$ such that $t \circ N = 0$ is true the statement $t \circ rad(A) = 0$ is valid. Thus, $C_T(N) = C_T(rad(A))$ is valid (One inclusion is always true.). The centralizer $C_T(rad(A))$ is – by using $A = rad(A) \oplus T$ and the commutativity of T – exactly $Z(A) \cap T$. Part (iii) is now a consequence of the theorems 25 and 20.

ad(iv): Part (iv) is proven by analogue argumentation as used in part (iii) and may be proven by the reader as an exercise (Theorem 14 is to be used, too.).◊

The previous results are mostly independent from a fixed radical complement:

Proposition 10 *Let K be a field possessing at least three elements, A a finite-dimensional associative solvable K-algebra possessing a separable factor algebra by its nilradical, $T, \hat{T}$ radical complements of $rad(A)$ in A and $\hat{C}$ a subalgebra of T. If $\hat{M}$ is a maximal Lie nilpotent subalgebra of $A°$ such that $\hat{C}$ is a radical complement of $rad(M)$ in M contained in $\hat{T}$, then an element $r \in rad(A)$ exists such that $\hat{M}^{1+r} = rad(M)^{1+r} \oplus \hat{C}^{1+r}$, $\hat{C}^{1+r} \subseteq C$ and $rad(M)^{1+r} \leq rad(A)$ are valid.*

A change of radical complement results in isomorphic copies of maximal Lie nilpotent subalgebras.

Proof. The proposition is a consequence of the solvability of A and the enhanced theorem of Wedderburn-Malcev (see [75]).◊

We close this section by bounding the Lie nilpotency class of maximal Lie nilpotent subalgebras. The extreme position of the Cartan subalgebras and of the nilradical appear again:

Corollary 3 *Let K be a field possessing at least three elements, A a finite-dimensional associative solvable K-algebra possessing a separable factor algebra by its nilradical, T an algebra complement of $rad(A)$ in A and M a maximal Lie nilpotent subalgebra of $A°$. The following statements are valid:*

(i) *The upper Lie central series of M is exactly $Z_n(M°) = Z_n(rad(M)°) \oplus VSEP(M)$ for all $n \in \mathbb{N}$. In particular, $cl(M°) = cl(rad(M)°) \leq cl(rad(M)) \leq cl(rad(A))$ is valid.*

(ii) $1 \leq cl(C_{rad(A)}(T)°) \leq cl(C_A(T)°) \leq cl(M°) \leq cl(nil(A°)) = cl(rad(A)°) \leq cl(rad(A))$.

Proof. ad(i): By using theorem 24 we conclude that $VSEP(M)$ is central in M and $M = rad(M) \oplus VSEP(M)$ is valid. Hence, the statement for the central chain in (i) is proven. The rest of part (i) is a consequence of the fact that the associative class of nilpotency is not smaller than the Lie class of nilpotency. By using the solvability of A we deduce $rad(M) \leq rad(A)$.

ad(ii): As proven in part (i) the statement $M = rad(M) \oplus VSEP(M)$ is valid. The fully separable part of M is situated – by using the theorems 14 and 20 – between the fully separable part of the nilradical and the Cartan subalgebra $C_A(T)$. The dual statement is valid by using proposition 9 for the nilpotent part $rad(M)$, and part(ii) is proven.⋄

We have proven diverse properties, characterizations and relationships for maximal Lie nilpotent subalgebras. Within the next section we focus on their absolute number and the number of isomorphism classes of these subalgebras. For this, we will prove that finite-dimensional separable commutative algebras – in our context radical complements – belong to the so-called class of futile algebras. Within the next section we summarize some basic facts of this class of algebras and analyze this class in more details.

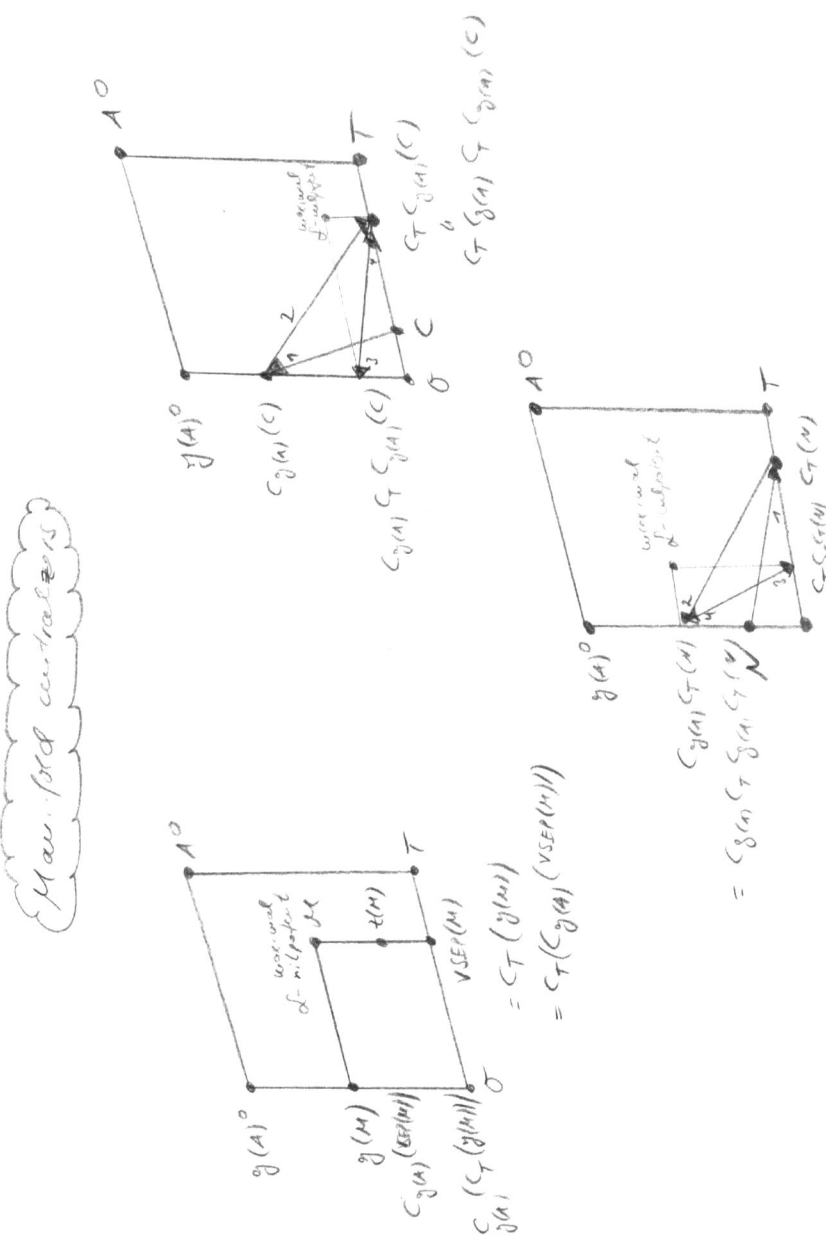

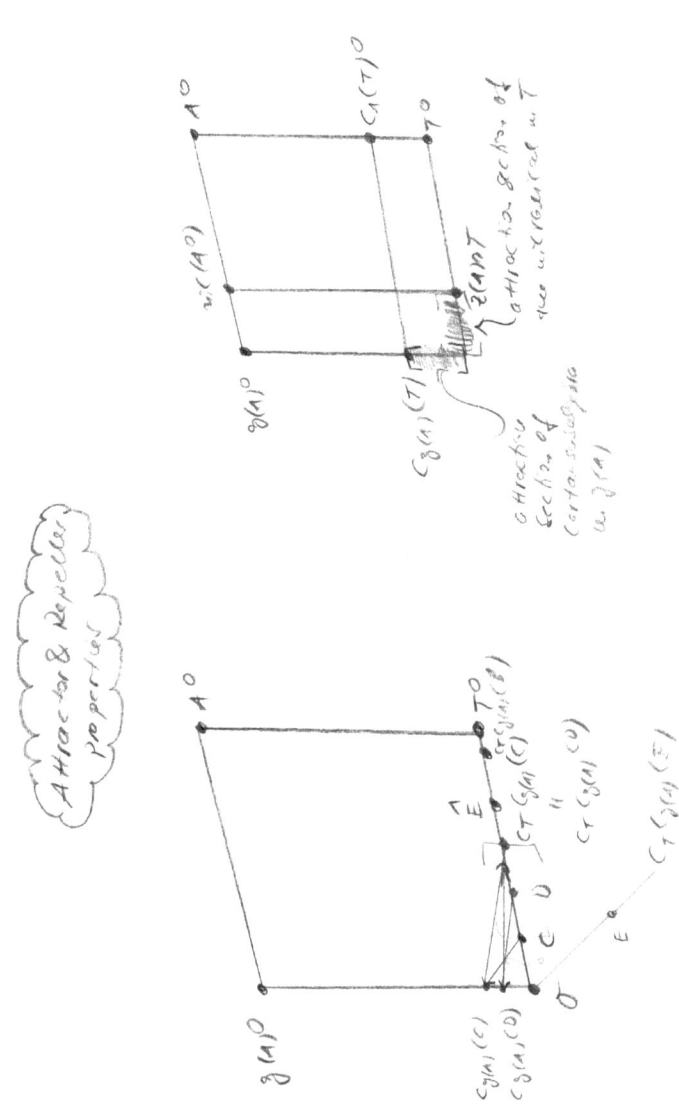

6.3 Futile algebras

6.3.1 Futility and radical complements

Definition 3 A K-algebra is called futile, if it possesses only finite many K-subalgebras.◇

Remark 6 The property which define futile algebras arise within Galois theory: the primitive element theorem analyzes on what terms a field extension is simple. Let $(K; L)$ be a field extension and $a \in L$. The field extension is called simple if $L = K(a)$ is valid. The element a is called primitive, and is not uniquely determined.

The primitive element theorem was proven by Evariste Galois and published 1829 by Niels Abel. Two theorems exist which are called the primitive element theorem. But the second one is a consequence of the first one:
A field extension $(K; L)$ is simple, if L is of the form $L = K(a, c_1, \cdots, c_n)$ such that a is algebraic and $c_1, \cdots, c_n$ are separable over K.
In particular, every finite-dimensional separable field extension is simple.

Within Galois theory all Galois field extensions are simple and every automorphism of the Galois group is determined by the action on the primitive element (see also [80]).

It can be proven – and this may be done by the reader within the exercises – that a finite-dimensional field extension is simple if and only if only finite many intermediate fields exist (see e.g. [84]). Thus, it is straightforward to prove that finite-dimensional simple field extensions possess only finite many unital subalgebras: because every unital subalgebra is a subfield based on proposition 5.

We are able to prove that that finite-dimensional simple field extension are indeed futile algebras. For this we use an argument stated in [40]. Let S be a subalgebra of the extension field different from zero. We choose an element $0 \neq s \in S$. On S we define the right multiplication with s which is the map $x \mapsto xs$. This function is injective because L is a division algebra and $s \neq 0$. By using the finite dimension of L we deduce that the map is surjective, too. Hence, an element $x \in S$ exists such that $xs = s$ is valid. Hence (because L is a division algebra and $s \neq 0$), $x = 1 \in S$ is valid.

Thus, we have proven that the subalgebras of L are either the zero-space or the finite many intermediate subfields.◇

Hisao Tominaga characterizes futile associative algebras over fields within his article [70]:

Theorem 26 *(H. Tominaga) Let A be an associative K-algebra based on a field K. The following statements are equivalent:*

(i) A is futile.

(ii) A possesses only finite many commutative subalgebras.

(iii) A is finite or the following statement is valid: A is finite-dimensional and every subalgebra is generated by a single element (monogenic).⋄

Michiel Kosters analyzes further properties of futile algebras further in his article [37]. We want to prove that commutative separable associative algebras are futile and, in addition, we want to estimate the number of its subalgebras. By using remark 6 we derive that a separable field extension is futile. Separable commutative finite-dimensional associative algebras are direct products of separable field extensions. Thus, it is natural to analyze direct products of futile algebras. The following lemma – which is the algebra version of the lemma of Jean-Baptiste Goursat[1] in group theory – it is possible to do this analysis. Michiel Kosters has used this lemma in [37], too. The lemma demonstrates the determination of subalgebras of direct products. Its proof is similar to the one in group theory and may be done by the reader as an exercise.

Lemma 12 *(Goursat) Let A, B be associative K-algebras. There is a bijective connection between 5-tuples (C, D, I, J, φ) – such that C is a K-subalgebra of A, D is a K-subalgebra of B, I is an ideal of C, J is an ideal of D and φ is a K-algebra isomorphisms between C/I and D/J – and K-subalgebras of $A \times B$ defined by:*

$$(C, D, I, J, \varphi) \mapsto \{(a; b) \mid (a; b) \in C \times D, \varphi(a + I) = b + J\}.$$⋄

---

[1]Édouard Jean-Baptiste Goursat (21 May 1858 to 25 November 1936) was a French mathematician, now remembered principally as an expositor for his Cours d'analyse mathématique, which appeared in the first decade of the twentieth century. It set a standard for the high-level teaching of mathematical analysis, especially complex analysis. This text was reviewed by William Fogg Osgood for the Bulletin of the American Mathematical Society. This led to its translation in English by Earle Raymond Hedrick published by Ginn and Company. Goursat also published texts on partial differential equations and hypergeometric series. Edouard Goursat was born in Lanzac, Lot. He was a graduate of the École Normale Supérieure, where he later taught and developed his Cours. At that time the topological foundations of complex analysis were still not clarified, with the Jordan curve theorem considered a challenge to mathematical rigour (as it would remain until L. E. J. Brouwer took in hand the approach from combinatorial topology). Goursat's work was considered by his contemporaries, including G. H. Hardy, to be exemplary in facing up to the difficulties inherent in stating the fundamental Cauchy integral theorem properly. For that reason it is sometimes called the Cauchy Goursat Theorem.

But we choose a different concept: we start the analysis for the algebra K^n ($n \in \mathbb{N}$, K a field) and transfer the results to commutative separable associative algebras.

We start our analysis by connecting unital and non-unital subalgebras. Unital subalgebras only exist if the algebras is unitary. In this case there are two types of subalgebras: unital and non-unital. For example, ideals are non-unital subalgebras. Later on we will use this lemma to prove a connection between A and A^K concerning futility.

Lemma 13 *Let A be an associative K-algebra and $(K;A)$ the adjunction of an unit of A. The map $T \mapsto (K;T)$ is a bijection between subalgebras of A and unital subalgebras of $(K;A)$.*

Proof. For non-empty sets C, D, E, F the statement $C \times D = E \times F$ is valid if and only if $C = E$ and $D = F$ are true. Hence, the map $T \mapsto (K;T)$ is injective. Let C be an unital subalgebra of $(K;A)$. $(K;0) \subseteq C$ is valid because $(1,0)$ is the unit element. We define $T := \{a \mid a \in A, \exists k_a \in K : (k_a;a) \in C\}$. We will prove that T is a subalgebra of A and $(K;T) = C$ is valid. Let $l \in K$ and $a, b \in T$. There exist $k_a, k_b \in K$ such that $(k_a;a), (k_b,b) \in C$ are valid. $k(k_a, a) = (kk_a, ka) \in C$ is true, and thus $ka \in T$ is valid. In addition, $(k_a, a) + (k_b, b) = (k_a + k_b; a + b) \in C$ is valid, and we derive $a + b \in T$. $(k_a, a)(k_b, b) = (k_a k_b; k_a b + k_b a + ab) \in C$ is true. Because of $(1,0) \in C$ we conclude $(k_a, 0) \in C$, and hence $(0, a) \in C$ is true. We derive $(0, k_b a) \in C$. By using an analogue argumentation we conclude $(0, k_a b) \in C$. The difference of $(k_a, a)(k_b, b)$ and $(0; k_a b + k_b a)$ is $(k_a k_b, ab)$, and thus its contained in C. We derive $ab \in T$. We have proven that T is a subalgebra of A, and is straightforward to derive $C \subseteq (K, T)$. If $(k, t) \in (K, T)$ is valid, then there exist $k_t \in K$ such that $(k_t, t) \in C$ is valid. We conclude $(k, t) = (k_t, t) - (k_t, 0) + (k, 0) \in C$ because C is unital and $(1;0)$ is the unit.⋄

Corollary 8 *Let K be a field and A a finite-dimensional associative semisimple K-algebra. A possesses as many subalgebras as unital subalgebras of $A \times K$ exist. In particular, K^n possesses as many subalgebras as unital subalgebras of K^{n+1} exist.*

Proof. By using theorems in [75] the adjunction by an unit A^K of A is isomorphic to $A \times K$ (because A is unitary). Lemma 13 finishes the proof.⋄

We will prove that the number of (unital) subalgebras of K^n is connected to the so-called Bell numbers. The n-th Bell number $B(n)$ is the number of unordered set partitions of a set consisting of n elements. Additional information concerning Bell numbers are included in [85] and on pages 84 and

85 in [56]. Bell numbers arise also within theorem 7 for the Solomon-Tits algebra as dimension of the factor algebra by the nilradical. Now we will link the Bell numbers to the number of (unital) subalgebras of K^n. It is well-known that the number of unordered set partitions of a set consisting of n elements is exactly the number of equivalence relations on the same set. The next lemma is based on exercise 8 on page 84 in [56]. If $a_1, \cdots, a_n$ are elements of a algebra and T a subset of $\underline{n}$, then we define $a_T := \sum_{t \in T} a_t$.

Lemma 14 *Let K be a field and $n \in \mathbb{N}$. K^n possesses exactly $B(n)$ unital subalgebras.*

Add-on: If $e_1, \cdots, e_n$ are the pairwise orthogonal primitive idempotents of K^n (these are the idempotents $e_i = (0, \cdots, 0, 1, 0, \cdots, 0)$, $i \in \underline{n}$), then the map

$$\mathcal{P} \mapsto \langle e_T \mid T \in \mathcal{P} \rangle_K$$

is a bijection between the partitions of $\underline{n}$ and the unital subalgebras of K^n.

Proof. For two sets T, S of $\underline{n}$ the statement $e_T = e_S$ is only for $T = S$ valid. Let $\mathcal{P}$ be a partition of $\underline{n}$. We define the subalgebra $B_\mathcal{P} := \langle e_T \mid T \in \mathcal{P} \rangle_K$. $\mathcal{P}$ is a partition of $\underline{n}$, and thus the elements $e_T, T \in \mathcal{P}$ are exactly the pairwise orthogonal primitive idempotents of $B_\mathcal{P}$ and their sum is the unit element of K^n. In particular, $B_\mathcal{P}$ is an unital subalgebra of K^n. If $\mathcal{Q}$ is another partition of $\underline{n}$ such that $B_\mathcal{P} = B_\mathcal{Q}$ is valid, then we conclude by the uniqueness of the primitive orthogonal idempotents that the sets $\{e_T \mid T \in \mathcal{P}\}$ and $\{e_T \mid T \in \mathcal{Q}\}$ are identical. We derive that both partitions are identical, too. Hence, the map $\mathcal{P} \mapsto B_\mathcal{P}$ is injective, and we prove that it is surjective, too. Let B be an unital subalgebra of K^n. By using theorems in [75] this subalgebra is diagonalizable, too, and thus it is isomorphic to $K^{dim_K(B)}$. We conclude that B is possessing a basis consistent of orthogonal idempotents $f_1, \cdots, f_{dim_K(B)}$ of K^n. For each of these idempotents f_j a subset S_j of $\underline{n}$ exists such that $f_j = e_{S_j}$ is valid, because every idempotent of K^n is the sum of some of the idempotents $e_1, \cdots, e_n$. The idempotents f_j are orthogonal, and thus the sets S_j are disjoint. Their union is $\underline{n}$ because the sum of these idempotents is the unit element K^n (B is unital). We derive that the sets S_j form a partition $\mathcal{Q}$ such that $B = \langle e_T \mid T \in \mathcal{Q} \rangle_K$ is valid.◇

Within the next lemma we determine the number of ideals of K^n based on a result in a more general context:

Lemma 15 *Let K be a field, A a finite-dimensional associative semisimple K-algebra and $A = \bigoplus_{i=1}^{n} J_i$ a direct decomposition in simple ideals of A. The following statements are valid:*

(i) A possesses exactly 2^n ideals.

(ii) A possesses at least 2^n idempotents.

(iii) A possesses exactly 2^n idempotents if and only if A is a basic algebra.

In particular, K^n possesses exactly 2^n ideals and idempotents.

Proof. ad(i): By using results in [40] every ideal is the sum of certain minimal ideals. Every sum of distinct minimal ideals results in a different ideal of A. Hence, the number of ideals of A is exactly the order of the power set of $\underline{n}$, and part (i) is proven.

ad(ii)+(iii): Every minimal ideal J_i possesses at least the idempotents 0 and 1_{J_i}. By these idempotents we can construct 2^n different idempotents. If every subalgebra J_i is a division algebra, then in every J_i only these two idempotents are existing. Every idempotent of A is a sum of idempotents of the J_i. If one of the J_i is no division algebra – and then a full matrix algebra based on a division algebra –, then more than two idempotents are existing in that particular J_i (e.g. the matrix possessing only the entry 1 in the first column and otherwise the entry 0). Thus, parts (ii) and (iii) are proven.⋄

We summarize our results about the number of certain substructures of K^n based on corollary 8, lemma 14 and lemma 15:

Theorem 27 *Let K be a field and $n \in \mathbb{N}$. The following statements are valid:*

(i) K^n is futile.

(ii) K^n possesses exactly $B(n+1)$ subalgebras.

(iii) K^n possesses exactly $B(n)$ unital subalgebras.

(iv) K^n possesses exactly 2^n ideals and 2^n idempotents.

(v) K^n possesses exactly $B(n+1) - B(n)$ non-unital subalgebras.

(vi) $B(n+1) \geq B(n) + 2^n - 1$⋄

We will use the following insight of separable algebras, and the proof can be studied by the reader (e.g. in [50]):

Proposition 11 *Let K be a field and A a finite-dimensional associative separable K-algebra. The L-Algebra $A \otimes L$ is semisimple for every field extension L of K. In particular, A possesses an extension field L of K such that the algebra $A \otimes L$ is isomorphic to a direct sum of full matrix-algebras over L. If A is commutative, then an extension field L of K exists such that the L-algebra $A \otimes L$ is isomorphic to $L^{dim_K(A)}$.*⋄

For commutative separable algebras we deduce the following corollary (see e.g. [56], pages 90 ff.):

Corollary 4 *Let K be a field, $L_1, \cdots, L_n$ separable extension fields of K and $A := \bigoplus_{i=1}^{n} L_i$. A is futile, and the number of all resp. all unital K-subalgebras of A is at least $B(dim_K(A) + 1)$ resp. $B(dim_K(A))$.*

Proof. We use proposition 11 and theorem 27. Let L be a splitting field for A. The L-algebra $A \otimes L$ is isomorphic to $L^{dim_K(A)}$. In addition, the map $T \mapsto T \otimes L$ is injective from the set of unital subalgebras of A into $A \otimes L$ because $(T \otimes L) \cap (A \otimes 1) = T \otimes 1$ is valid. $T \otimes 1 = S \otimes 1$ is true if and only if $T = S$ is valid for all subalgebras T, S of A. Hence, the proof is finished.⋄

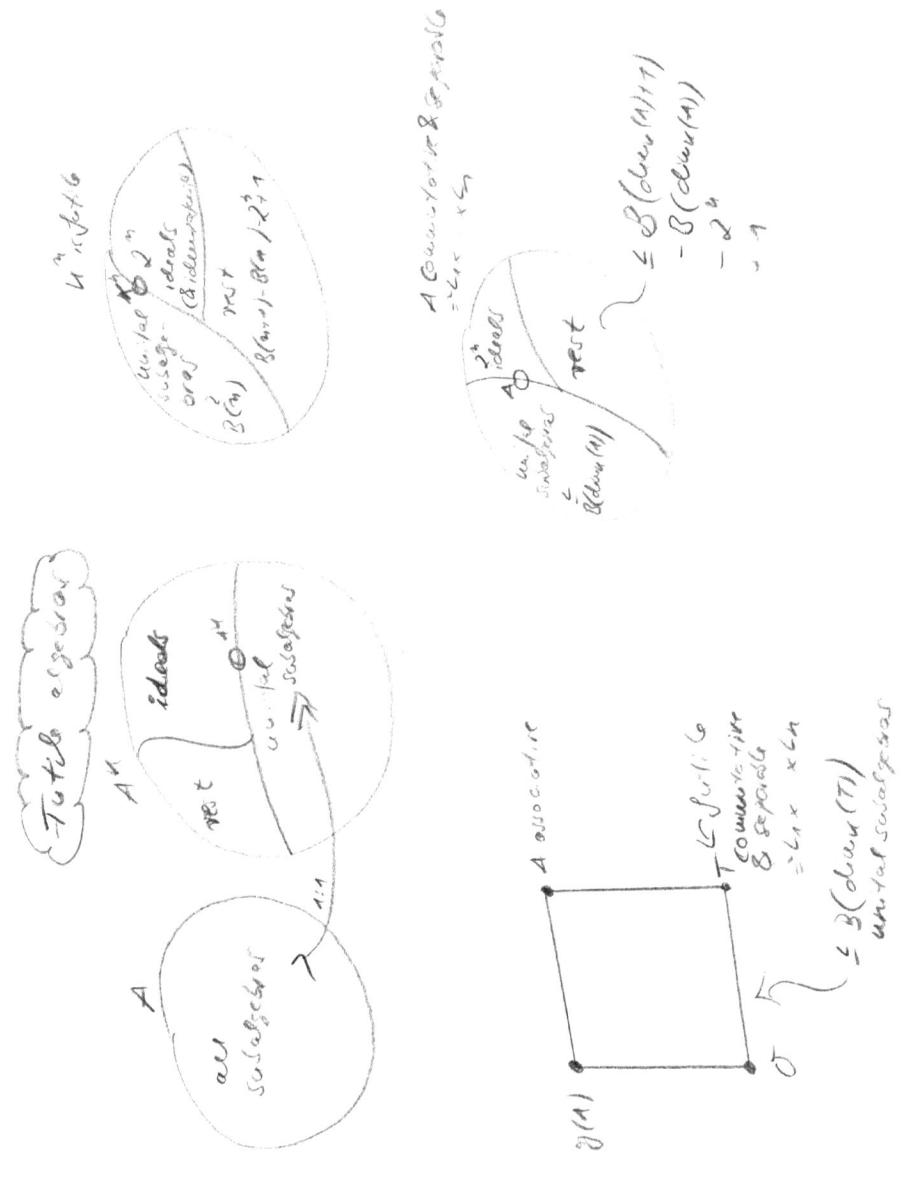

6.3.2 Futility and unital subalgebras

The focus of the last section was to deduce that commutative separable algebras are futile and to bound the number of their unital subalgebras. In our context of maximal nilpotent Lie subalgebras we will use this result for bounding the number of isomorphism classes of maximal Lie nilpotent subalgebras based on solvable algebras.

Within this section – which is an excursus – we focus on a connection between non-unital and unital subalgebras for unitary algebras and deduce a characterization of futile algebras based on this connection. The main tool for our analysis is the Chinese Remainder Theorem for K-algebras. In the literature this theorem is proven only for rings. Thus we state a proof for associative algebras, too.

Definition and remark 4 Let A be an associative K-algebra, $n \in \mathbb{N}$, I_i K-ideals of A for all $i \in \underline{n}$. The sum resp. the intersection resp. the product of two ideals I_1 and I_2 is defined by $I_1 + I_2 := \{i_1 + i_2 \mid (i_1; i_2) \in I_1 \times I_2\}$ resp. $I_1 \cap I_2 := \{x \mid x \in I_1, x \in I_2\}$ resp. $I_1 \cdot I_2 := \langle i_1 \cdot i_2 \mid (i_1; i_2) \in I_1 \times I_2 \rangle_K$. All three sets are K-ideals of A. The n-fold sum resp. intersection resp. product of n ideals $I_1, \ldots, I_n$ is defined recursively. Again, all three substructures are K-ideals of A. The n-fold product of I_1 with itself is defined by $(I_1)^n := I_1 \cdot \cdots \cdot I_1$. We remark that $I_1 + \cdots + I_n = \langle i_1 + \cdots + i_n \mid (i_1; \ldots; i_n) \in I_1 \times \cdots \times I_n \rangle_K$ and $I_1 \cdot \cdots \cdot I_n \subseteq I_1 \cap \cdots \cap I_n$ are valid. The distributive inclusion $(I_1 + I_3) \cdot (I_2 + I_3) \subseteq I_1 \cdot I_2 + I_3$ is true. I_1 and I_2 are called comaximal if $A = I_1 + I_2$ is valid.

Proposition 12 Let A be an associative unitary K-algebra, $n \in \mathbb{N}$ and $I_1, \cdots, I_n$ pairwise comaximal K-ideals of A. The K-ideals $I_1 \cap \cdots \cap I_{n-1}$ and I_n are comaximal.

Proof. $I_1, \cdots, I_n$ are pairwise comaximal K-ideals of A, and thus $A = I_i + I_n$ is valid for all $i \in \underline{n-1}$. A is unitary, and we deduce $A = A^{n-1} = (I_1 + I_n) \cdot \cdots \cdot (I_{n-1} + I_n)$. A straightforward calculation (using an induction argument, the ideal properties of $I_1, \cdots, I_n$ and definition and remark 4) shows us that this product is contained in $I_1 \cdot \cdots \cdot I_{n-1} + I_n$. The proof is finished by using definition and remark 4. ⋄

The next proposition is straightforward to prove and may be done by the reader within the exercises:

Proposition 13 Let A be an associative (unitary) K-algebra, $n \in \mathbb{N}$ and $I_1, \cdots, I_n$ K-ideals of A. The map

$$\chi : A/(\bigcap_{i=1}^{n} I_i) \longrightarrow \bigoplus_{i=1}^{n}(A/I_i), a + (\bigcap_{i=1}^{n} I_i) \mapsto (a + I_1; \cdots ; a + I_n)$$

is a(n) (unitary) K-algebra monomorphism.◇

Lemma 16 *(Chinese Remainder Theorem for associative algebras)* Let A be an associative unitary K-algebra, $n \in \mathbb{N}$ and $I_1, \cdots I_n$ pairwise comaximal K-ideals of A. The map

$$\chi : A/(\bigcap_{i=1}^{n} I_i) \longrightarrow \bigoplus_{i=1}^{n}(A/I_i), a + (\bigcap_{i=1}^{n} I_i) \mapsto (a + I_1; \cdots ; a + I_n)$$

is a(n) (unitary) K-algebra isomorphism.◇

Proof. Based on proposition 13 we need to show that χ is surjective. This is done by an induction argument. The case $n = 1$ is true. We proceed with the case $n = 2$. Let I_1, I_2 be two comaximal ideals. Let $x, y \in A = I_1 + I_2$. Thus, there exists $i_1, j_1 \in I_1$ and $i_2, j_2 \in I_2$ such that $x = i_1 + j_1$ and $y = i_2 + j_2$ are valid. It is straightforward to verify that $((i_2 + j_1) + I_1 \cap I_2)\chi = (x + I_1; y + I_2)$ is true. The general case is proven using the result for the case $n = 2$ and by an induction argument based on proposition 12. The K-ideals $I_1 \cap \cdots \cap I_{n-1}$ and I_n are comaximal by using the stated proposition. We deduce that $A/(\bigcap_{i=1}^{n} I_i)$ is isomorphic to $(A/(\bigcap_{i=1}^{n-1} I_i)) \oplus (A/I_n)$. Now we can apply induction on $A/(\bigcap_{i=1}^{n-1} I_i)$.◇

A consequence of the Chinese Remainder Theorem is the following corollary:

Corollary 9 Let A be an associative finite-dimensional unitary K-algebra, $n \in \mathbb{N}$ and $I_1, \cdots, I_n$ pairwise disjoint K-ideals of A of codimension 1. The K-dimension of $A/(\bigcap_{i=1}^{n} I_i)$ is exactly n.◇

Definition 4 Let A be an associative unitary K-algebra. A is called 1-futile if A possesses only finite many unital subalgebras. A is called non-1-futile if A possesses only finite many non-unital subalgebras. A is called nilpotent-futile if A possesses only finite many nilpotent subalgebras.

Theorem 28 Let K be a field and A a finite-dimensional associative unitary K-algebra. The following statements are valid:

(i) A is futile.

(ii) A is 1-futile.

In particulary, if A is 1-futile, then A is nilpotent-futile and non-1-futile.

Proof. Let A be 1-futile. Let S be a non-unital subalgebra of A. We define $S^K := S \oplus K1_A$. It is straightforward to prove that S^K is a unital subalgebra of A and S is an K-ideal of S^K. We focus on the map $^K : S \mapsto S^K$ from the set $\mathfrak{T}_{non\,1}$ of non-unital to the set $\mathfrak{T}_1$ of unital subalgebras of A. Because of the finiteness of $\mathfrak{T}_1$ the image of ^{K} is finite, too. $\mathfrak{T}_{non\,1}$ is infinite, and thus there is one subalgebra $S \oplus K1_A$ – based on a non-unital subalgebra S – for which there are infinite many pre-images under ^{K}: the set $\{T \mid T \in \mathfrak{T}_{non\,1}, T \oplus K1_A = S \oplus K1_A\}$ is infinite. A is finite-dimensional, and thus $S \oplus K1_A$ is finite-dimensional, too. Let $n \in \mathbb{N}$ such that $n = dim_K(S \oplus K1_A)$ is valid. $\{T \mid T \in \mathfrak{T}_{non\,1}, T \oplus K1_A = S \oplus K1_A\}$ is infinite, and thus there are pairwise disjoint non-unital subalgebras $T_1, \cdots, T_{n+1}$ such that $T_i \oplus K1_A = S \oplus K1_A$ is true for all $i \in \overline{n+1}$. All $n+1$ subalgebras T_i are K-ideals of $S \oplus K1_A$ of codimension 1. These K-ideals satisfy the assumptions of corollary 9 with $A = S \oplus K1_A$. Thus we deduce that the dimension of $(S \oplus K1_A)/(\bigcap_{i=1}^{n+1} T_i)$ is exactly $n+1$. But this a contradiction to the fact that $n = dim_K(S \oplus K1_A)$ is true. ◇

A direct consequence of this theorem and lemma 13 is:

Corollary 10 *Let K be a field and A a finite-dimensional associative K-algebra. The following statements are valid:*

(i) A is futile.

(ii) A^K is futile. ◇

Remark 7 (i) We focus on the map ^{K} used within theorem 28 from the set $\mathfrak{T}_{non\,1}$ of non-unital to the set $\mathfrak{T}_1$ of unital subalgebras. We have proven that the finiteness of $\mathfrak{T}_1$ lets us deduce that $\mathfrak{T}_{non\,1}$ is finite, too, using the map ^{K} and the finite dimension of the underlying algebra. We state two examples such that the map is – within the first example – injective but not surjective and – within the second example – not injective but surjective. The first example focusses on the $\mathbb{R}$-algebra $\mathbb{C}$ which has dimension 2. Based on definition and remark 6 there are two unital subalgebras – which are $\mathbb{C}$ and $\mathbb{R}$. The only non-unital subalgebra is the zero space. Thus the map $^\mathbb{R}$ has the desired characteristics. The second example deals with the three-dimensional $\mathbb{R}$-algebra $\mathbb{R}^3$. Based on theorem 27 this algebra contains $B(3) = 5$ unital subalgebras and $B(4) - B(3) = 15 - 5 = 10 > 5 = B(3)$ non-unital subalgebras. The following pictures illustrates the action of $^\mathbb{R}$ and shows that the map is surjective but not injective.

Properties of $\oplus: \mathbb{R}^1 \to \mathbb{R}^3$

$\mathbb{R}^3 = \langle e_1, e_2, e_3 \rangle_\mathbb{R}$ pairwise orthogonal, idempotent 1

$\langle e_1, e_2, e_3 \rangle_\mathbb{R} \quad \oplus \mathbb{R}^1$
$\langle e_1, e_3 \rangle_\mathbb{R} \quad \oplus \mathbb{R}^1$
$\langle e_1, e_2 \rangle_\mathbb{R}$

$\langle e_1, e_2 + e_3 \rangle_\mathbb{R}$
$\langle e_2, e_3 \rangle_\mathbb{R}$
$\langle e_1 + e_2 \rangle_\mathbb{R}$
$\langle e_3 \rangle_\mathbb{R}$

$\langle e_1 + e_2 + e_3 \rangle_\mathbb{R} = \mathbb{R} \cdot 1$

$\langle e_1 \rangle_\mathbb{R} \quad \langle e_2 \rangle_\mathbb{R} \quad \langle e_3 \rangle_\mathbb{R}$

0

① $\mathbb{R}^1 \to$ surjective but not injective
○ $\mathbb{R}^1 \to$ injective but not surjective

- vertical subalgebras
○ non-vertical subalgebras

$B(3) = 5$
$B(4) = 15$
$B(4) - B(3) = 10$

① $\oplus \mathbb{R} \to \mathbb{R}^1$
① $\oplus \mathbb{R}^1 \to$ injective but not surjective

(ii) The aim of this part is to deduce a connection between nilpotent-futile and 1-futile algebras. Let A be a 1-futile finite-dimensional associative unitary algebra. Based on theorem 28 we know already that A is nilpotent-futile. In this part we deduce that the number of nilpotent subalgebras is not greater than the number of unital subalgebras. Let S, T be subalgebras of A such that $S^K = S \oplus K1_A = T \oplus K1_A = T^K$ is valid. Hence, S and T are of equal dimension. It is straightforward to prove that S^K is isomorphic to the adjunction of a unit of S. Within [75] it is proven that the nilradical of S^K is exactly the nilradical of S. We deduce $rad(S) = rad(S^K) = rad(T^K) = rad(T)$. If S or T is nilpotent, then we deduce $S = rad(S) = rad(T) \leq T$ or $T = rad(T) = rad(S) \leq S$ is valid. Comparing dimension we conclude $S = T$. In particular, ^{K} is injective on the set of nilpotent subalgebras.

(iii) The aim of this part is to deduce that non-1-futile algebras are not necessarily futile. Our example is based on definition and remark 6. We focus on a non-simple finite-dimensional field extension $(K; L)$. We have proven that the zero-space is the only non-unital subalgebra. In addition, there are infinite many intermediate fields which are unital subalgebras. Thus, we need an example of a non-simple finite-dimensional field extension. The following example is taken from mathstackexchange [87]:

> Let $F = \mathbb{F}_p(X, Y)$ be the field of rational functions in variables X, Y over the finite field of p elements. Let $K = \mathbb{F}_p(X^p, Y^p)$ be a subfield. Note that for any $f \in F$, $f^p \in K$. Deduce from this that F/K is not a simple extension. First of all $\mathbb{F}_p(X^p, Y^p)$ is a function field, which means that we may treat X^p, Y^p as variables. The polynomial $(T - X)^p = T^p - X^p$ is irreducible over $\mathbb{F}_p(Y^p)[X^p]$ by Eisenstein's criterion, but then also over the field of fractions $\mathbb{F}_p(X^p, Y^p)$. Hence, $\mathbb{F}_p(X, Y^p)$ has degree p over $\mathbb{F}_p(X^p, Y^p)$. Similarily one proves that $\mathbb{F}_p(X, Y)$ has degree p over $\mathbb{F}_p(X, Y^p)$. Hence, $F = \mathbb{F}_p(X, Y)$ has degree p^2 over $K = \mathbb{F}_p(X^p, Y^p)$. But since $F^p \subseteq K$, every element of F has degree $\leq p$ over K, so that it cannot generate F/K.⋄

6.4 Finiteness of the number of isomorphism classes

In this section we prove that only finite many isomorphism types of maximal Lie nilpotent subalgebras exist. For this, we use the results for futile algebras as well as the results for the determination and characterization of all maximal nilpotent Lie subalgebras. Within theorems 20 and 14 it is proven that there is exactly one class of isomorphism for the nilradical and for the Cartan subalgebras.

Theorem 29 *Let K be a field possessing at least three elements, A a finite-dimensional associative solvable K-algebra possessing a separable factor algebra by its nilradical, T a radical complement of $rad(A)$ in A, $\mathcal{M}(T) := \{C \mid C \leq T, C = C_T(C_{rad(A)}(C))\}$ and $m_T := \mid \mathcal{M}(T) \mid$. The following statements are valid:*

(i) The nilradical of A° is unique.

(ii) There is exactly one class of isomorphism of Cartan subalgebras of A°.

(iii) There are only finite many classes of isomorphism of maximal Lie nilpotent subalgebras of A°.

(iv) An upper bound for the number of classes of isomorphism of maximal Lie nilpotent subalgebras of A° is given by $m_T \leq B(dim_K(T))$.

(v) If T is isomorphic to K^n for a natural number $n \in \mathbb{N}$, then $m_T \leq B(n)$ is valid.

Proof. ad(i) and (ii): see theorems 20 and 14.

ad(iii)-(v): Within proposition 10 we have proven that all maximal Lie nilpotent subalgebras are determined – up to conjugation within $1 + rad(A)$ – in a way that the fully separable part is contained in one fixed radical complement T. By using the theorems 24 and 25 we have derived that all fully separable parts in T can be described by the set $\mathcal{M}(T)$. If the fully separable part is fixed, then the nilpotent part is fixed, too (again by theorems 24 and 25). Thus the whole maximal Lie nilpotent subalgebra is fixed (see again theorems 24 and 25). Hence, the number of maximal Lie nilpotent subalgebras for which the fully separable part is describable by the set $\mathcal{M}(T)$ is at least the order of $\mathcal{M}(T)$. The set $\mathcal{M}(T)$ consists of unital subalgebras of T (because $1_A \in C_T(C_{rad(A)}(C)) = C$), and by using corollary 4 and the futility of T the set $\mathcal{M}(T)$ is finite. By using the same theorems the order of the set $\mathcal{M}(T)$ can be estimated as stated by $B(dim_K(T))$.◇

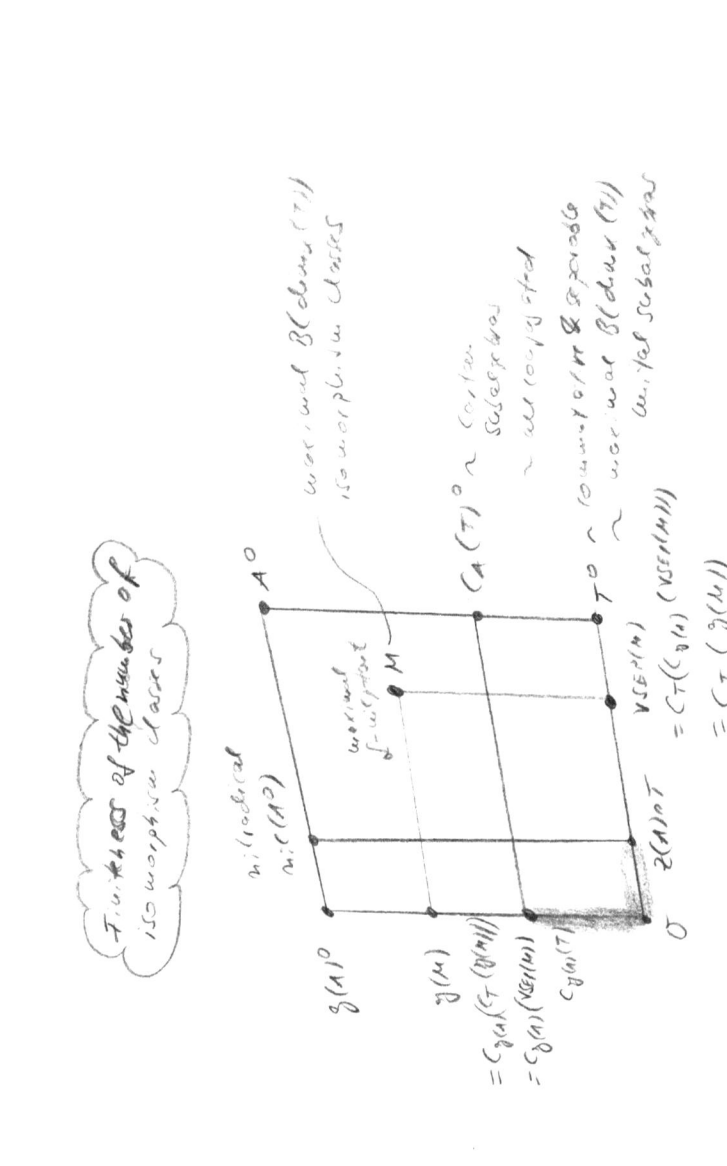

6.5 Cardinalities

In this section we analyze connections between sets (and their cardinalities) of subalgebras which arise within the determination of maximal Lie nilpotent subalgebras. We begin to define these sets.

Definition and remark 5 Let K be a field, A an associative finite-dimensional K-algebra possessing a separable factor algebra by its nilradical, and let T be a radical complement of $rad(A)$ in A. If M is a subalgebra of A, then $VSEP(M)$ is the set of all fully-separable elements of M. Within the analysis of Lie nilpotent associative algebras M we have proven that $VSEP(M)$ is exactly the (central) radical complement of M. Now we define sets of subalgebras based on special centralizer and double-centralizer properties:

- $\mathcal{A}_{DT} := \{C \mid C \leq T, C_T(C_{rad(A)}(C)) = C\}$
- $\mathcal{A}_{DJ} := \{C \mid C \leq rad(A), C_{rad(A)}(C_T(C)) = C\}$ and
- $\mathcal{A}_{MT} := \{M \mid M \leq A, M = rad(M) \oplus VSEP(M), C_T(rad(M)) = VSEP(M), C_{rad(A)}(VSEP(M)) = rad(M)\}$.

In addition, let $\mathcal{A}_M$ be the set of all maximal Lie nilpotent subalgebras of A°. We have proven that we can construct members of the third set $\mathcal{A}_{MT}$ by using members of the first two sets $\mathcal{A}_{DJ}$ and $\mathcal{A}_{DT}$. The next theorems demonstrate that this construction is complete. For this, we will use that all (separable) radical complements are conjugated under $1_A + rad(A)$ by the theorem of Wedderburn-Malcev.⋄

Theorem 30 *Let K be a field possessing at least three elements, A a finite-dimensional associative solvable unitary K-algebra possessing a separable factor algebra by its nilradical, T, S radical complements of $rad(A)$ in A and $r \in rad(A)$ such that $S = T^{1+r}$ is valid. The following statements are true:*

(i) *$\mathcal{A}_{MT}$ is exactly the set of all maximal Lie nilpotent subalgebras M of A° such that their fully-separable part $VSEP(M)$ is contained in T.*

(ii) *$\mathcal{A}_{MT}$ and $\mathcal{A}_{MS}$ are finite sets of the same cardinality. More precisely, the map $\mathcal{A}_{MT} \longrightarrow \mathcal{A}_{MS}, M \mapsto M^{1+r}$ is a bijection.*

(iii) *$\mathcal{A}_{DJ}$ and $\mathcal{A}_{MT}$ are finite sets of the same cardinality. More precisely, the map $\mathcal{A}_{DJ} \longrightarrow \mathcal{A}_{MT}, N \mapsto C_T(N) \oplus N$ is a bijection, and $M \mapsto rad(M)$ is its inverse map.*

(iv) *$\mathcal{A}_{DT}$ and $\mathcal{A}_{MT}$ are finite sets of the same cardinality. More precisely, the map $\mathcal{A}_{DT} \longrightarrow \mathcal{A}_{MT}, C \mapsto C_{rad(A)}(C) \oplus C$ is a bijection, and its inverse map is $M \mapsto VSEP(M)$.*

(v) $\mathcal{A}_{DT}$ and $\mathcal{A}_{DS}$ are finite sets of the same cardinality. More precisely, the map $\mathcal{A}_{DT} \longrightarrow \mathcal{A}_{DS}, C \mapsto C^{1+r}$ is a bijection.

(vi) The cardinality of $\mathcal{A}_{MT}$ can be estimated by the upper bound $B(dim_K(T))$.

(vii) $\mathcal{A}_M = \bigcup_{r \in rad(A)} (\mathcal{A}_{MT})^{1+r}$

Proof. ad(i): This statement is valid by using lemma 9.

ad(ii)+(v): These statements are derivable by using proposition 10.

ad(iii)+(iv): Both maps are well-defined by using theorem 25. Lemma 9 lets us derive that the nilradical resp. the central radical complement of $C_T(N) \oplus N$ is exactly N resp. $C_T(N)$. Analogously, the nilradical resp. the central radical complement of $C_{rad(A)}(C) \oplus C$ is exactly $C_{rad(A)}(C)$ resp. C. Hence, the composition of the stated maps is the identity map, and, in particular, the first map is injective. The surjectivity of the map is a consequence of theorem 25.

ad(vi): This statement is a consequence of theorem 29.

ad(vii): By using part (i) the stated union is contained in $\mathcal{A}_M$. If M is a maximal Lie nilpotent subalgebra, the lemma 9 ensures the existence of a radical complement containing the fully separable part of M. Thus, part (vii) is consequence of part (ii).$\diamond$

Remark 8 We remark the following concerning theorem 30:

(i): If the nilradical is infinite, then $\mathcal{A}_M = \bigcup_{r \in rad(A)} (\mathcal{A}_{MT})^{1+r}$ needs not to be finite. We have proven that each set of the union is finite and of the same cardinality. If K is finite, then – by using theorem 30 – the order of $\mathcal{A}_M$ can be estimated by the upper bound $\mid rad(A) \mid \cdot \mid \mathcal{A}_{MT} \mid$. For this value the number $\mid rad(A) \mid \cdot B(dim_K(T))$ is an upper bound.

(ii): The intersection of the sets $\mathcal{A}_{MT}$ – such that T varies over all radical complements – are not known to the author. One known aspect is the following one: If $r \in rad(A)$ and M is a maximal Lie nilpotent subalgebra such that $M \in \mathcal{A}_{MT} \cap \mathcal{A}_{MT}^{1+r}$ is valid, then the condition $r \in C_{rad(A)}(VSEP(M)) = rad(M)$ is true. From this we derive $M = M^{1+r}$. If we want to construct potentially new maximal Lie nilpotent subalgebras by passing from M to a conjugate of M, then the conjugator must be chosen outside of the nilradical of M.

The proof of part (i) and (ii) may be done by the reader as exercises.$\diamond$

⟨ Catavalves ⟩

$g(A)$ —————————— $|A_{MT}| \xleftarrow{\text{LB}(l_{\text{win}}(T))}$
         A
         •——————————————•
         | |
$N \overset{?}{\to} C_{g(N)}(N)$ 1:1   worry   1:1
$N \overset{?}{\to} C_{g(N)}(N)$ of $l_{in}\bar{g}_{ja}$
Subalgebras with   to $\uparrow \alpha$
double- iso to C_{res} $\uparrow \circlearrowleft$ $C = C_T \circ g(n)(C)$
property in $g(A)$ Subalgebras with
                                          double-center to C_{res}
         •——————————————•   property in T
         J   T

$|d_{og}| = |A_{MT}|$ $|d_{oT}| = |d_{MT}|$

6.6 Examples

Our standard examples – the Solomon-Tits algebras, the Solomon algebras, the algebras of triangular matrices and the solvable group algebras – will be analyzed in details separately in series III. For illustrating our results we focus on the triangular matrices in small dimension.

Within the following examples let K be a field possessing at least three elements, A a finite-dimensional associative solvable K-algebra possessing a separable factor algebra by its nilradical, T, S radical complements of $rad(A)$ in A and $r \in rad(A)$ mit $S = T^{1+r}$.

Example 4 *(maximal subalgebras)* If C is a maximal subalgebra of T and N a maximal subalgebra of $rad(A)$, then

$$C \leq C_T(C_{rad(A)}(C)) \leq T \text{ and } N \leq C_{rad(A)}(C_T(N)) \leq rad(A)$$

are valid. By the maximality of the subalgebras we derive that C and N are fixed under the double-centralizing or the double-centralizing yields to T or $rad(A)$. In the first case $C \oplus C_{rad(A)}(C)$ resp. $N \oplus C_T(N)$ is maximal Lie nilpotent by using theorem 25. In the other case the double-centralizing yields to a Cartan subalgebra resp. to the Lie nilradical by using theorems 14 and 20.⋄

Example 5 *(minimal non-nilpotent)* Let A° be minimal non-nilpotent. We prove that the Lie nilradical and the Cartan subalgebras are all maximal Lie nilpotent subalgebras. Let $M = rad(M) \oplus VSEP(M)$ be a maximal Lie nilpotent subalgebra of A°. We focus on the subalgebra $B := rad(A) \oplus VSEP(M)$. If $B = A$ is valid, then $VSEP(M)$ is a radical complement. In this case we derive a Cartan subalgebra (see theorem 14 and lemma 9). In the other case B is Lie nilpotent containing M. Thus, M and B are identical, and $rad(A) = rad(B) = rad(M)$ is valid. In this case we have derived the Lie nilradical (see theorem 20 and Lemma 9).

In the article [66] all minimal non-nilpotent solvable Lie algebras are classified by E. Stitzinger.⋄

Example 6 *(triangular matrices I)* We focus on the subalgebra of lower triangular matrices $\delta_{u,3}$ and analyze the powers of the nilradical. For this, we use lemma 10 and corollary 7. The nilradical yields by double-centralizing to the Lie nilradical and the zero-space to a Cartan subalgebra.
Now let us focus on the ideal $rad(A)^2$ which can be represented by the matrices of the form $\begin{pmatrix} 0 & 0 & 0 \\ 0 & 0 & 0 \\ a & 0 & 0 \end{pmatrix}$. $C_{D(3,K)}(rad(A)^2)$ is represented by the

matrices of the form $\begin{pmatrix} x & 0 & 0 \\ 0 & y & 0 \\ 0 & 0 & x \end{pmatrix}$. $C_{rad(A)}(C_{D(3,K)}(rad(A)^2))$ is exactly $rad(A)^2$. Thus, $rad(A)^2 \oplus C_{D(3,K)}(rad(A)^2)$ is another maximal Lie nilpotent subalgebra of $\delta_{u,3}{}^\circ$ representable by $\begin{pmatrix} x & 0 & 0 \\ 0 & y & 0 \\ a & 0 & x \end{pmatrix}$. (The calculations based on matrices are straightforward and may be done by the reader as exercises.) The next example is the dual version of this example.◇

Example 7 *(triangular matrices II)* We focus again on the subalgebra of lower triangular matrices $\delta_{u,3}$ and analyze all unital subalgebras of $D(3, K)$. By using lemma 10, corollary 7, lemma 14 and theorem 29 we can bound the number of isomorphic classes of maximal Lie nilpotent subalgebras. There are 5 unital subalgebras T of $D(3, K)$. For each such subalgebra we calculate the double-centralizing and determine a maximal Lie nilpotent subalgebra. By this, we determine 5 maximal Lie nilpotent subalgebras, and these are the potential isomorphic classes of maximal Lie nilpotent subalgebras. (The calculations based on matrices are straightforward and may be done by the reader as exercises.) Let e_1, e_2, e_3 be the orthogonal primitive idempotents of $D(3, K)$.

$D(3, K)$ is the unique 3-dimensional unital subalgebra of $D(3, K)$. Double-centralizing yields to a Cartan subalgebra:

$$\begin{pmatrix} 0 & 0 & 0 \\ 0 & 0 & 0 \\ 0 & 0 & 0 \end{pmatrix} \oplus \begin{pmatrix} a & 0 & 0 \\ 0 & b & 0 \\ 0 & 0 & c \end{pmatrix}.$$

The other extreme is given by the K-subspace generated by the unit (which is $e_1 + e_2 + e_3$). The double-centralizing yields to the nilradical:

$$\begin{pmatrix} 0 & 0 & 0 \\ a & 0 & 0 \\ c & b & 0 \end{pmatrix} \oplus \begin{pmatrix} d & 0 & 0 \\ 0 & d & 0 \\ 0 & 0 & d \end{pmatrix}.$$

Now we have to focus on the 2-dimensional unital subalgebras of $D(3, K)$ which are $T_1 := \langle e_1 + e_2, e_3 \rangle_K$, $T_2 := \langle e_1 + e_3, e_2 \rangle_K$ and $T_3 := \langle e_1, e_2 + e_3 \rangle_K$. It is straightforward to calculate that for all $i \in \underline{3}$ the condition $T_i := C_{D(3,K)}(C_{rad(\delta_{u,3})}(T_i)) = T_i$ is valid. The double-centralizing is stable from the first step. Thus we derive the following maximal Lie nilpotent subalgebras represented by matrices:

$$\begin{pmatrix} 0 & 0 & 0 \\ a & 0 & 0 \\ 0 & 0 & 0 \end{pmatrix} \oplus \begin{pmatrix} b & 0 & 0 \\ 0 & b & 0 \\ 0 & 0 & c \end{pmatrix} \text{ (for } T_1\text{)},$$

$$\begin{pmatrix} 0 & 0 & 0 \\ 0 & 0 & 0 \\ a & 0 & 0 \end{pmatrix} \oplus \begin{pmatrix} b & 0 & 0 \\ 0 & c & 0 \\ 0 & 0 & b \end{pmatrix} \text{ (for } T_2\text{) and}$$

$$\begin{pmatrix} 0 & 0 & 0 \\ 0 & 0 & 0 \\ 0 & a & 0 \end{pmatrix} \oplus \begin{pmatrix} b & 0 & 0 \\ 0 & c & 0 \\ 0 & 0 & c \end{pmatrix} \text{ (for } T_3\text{).} \diamond$$

A subalgebra which is not fixed by the double-centralizing is e.g. an ideal of the radical complement $D(3, K)$ because ideals are not unital. This fact (for ideals) is true in a wider context, and the reader may calculate the double-centralizing for all ideals of $\delta_{u,3}$ within the exercises.$\diamond$

Example 8 *(attraction section of the intersection)* We focus again on the algebra of lower triangular matrices $\delta_{u,4}$ based on a field K. Let e_1, e_2, e_3, e_4 be the orthogonal indecomposable idempotents. In view of lemma 14 we focus on the following partitions of $\underline{4}$: $P := \{\{1,2\},\{3\},\{4\}\}$, $Q := \{\{1,2\},\{3,4\}\}$ and $R := \{\{1,2,3\},\{4\}\}$. Let $D_X := \langle e_T \mid T \in X \rangle_K$ for all $X \in \{P, Q, R\}$, $D := D(4, K)$ and $J := rad(\delta_{u,4})$. A straightforward calculation lets us deduce that the identities $C_D(C_J(D_P)) = D_P = C_D(C_J(D_Q)) = C_D(C_J(D_R))$ are valid. D_P is three-dimensional. The intersection of D_Q and D_R is the K-subspace spanned by the unit element. Thus, the intersection is not within the same attraction section as for D_Q and D_R (because the intersection is contained in the attraction section of the whole radical complement, see also corollary 6).

6.7 Summary

We summarize our results concerning maximal Lie nilpotent subalgebras. The summary is also used to transfer the main results in the chapter after next to maximal nilpotent subgroups of the group of units.

Main theorem 3 *Let K be a field possessing at least three elements, A a finite-dimensional associative solvable unitary K-algebra possessing a separable factor algebra by its nilradical and T an algebra complement of $rad(A)$ in A. The following statements are valid:*

(i) *Every maximal Lie nilpotent subalgebra M is an associative unital subalgebra of the form $M = rad(M) \oplus VSEP(M)$ such that $VSEP(M)$ is central in M and $rad(M) \leq rad(A)$ is valid. Modulo conjugation* by a suitable element $1+r, r \in rad(A)$ the subalgebra $VSEP(M)$ is a subalgebra of T.

(ii) If $M = rad(M) \oplus VSEP(M)$ is a Lie nilpotent subalgebra possessing a central radical complement $VSEP(M)$ in T, then M is maximal Lie nilpotent if and only if

$$rad(M) = C_{rad(A)}(VSEP(M))) \text{ and } VSEP(M) = C_T(rad(M))$$

are valid.

(iii) All maximal Lie nilpotent subalgebras possessing a fully separable part contained in T are determinable by double-centralizing of unital subalgebras C of T resp. N of $rad(A)$: $C_T(C_{rad(A)}(C)) \oplus C_{rad(A)}(C_T(C_{rad(A)}(C)))$ and $C_{rad(A)}(C_T(N)) \oplus C_T(C_{rad(A)}(C_T(N)))$ are maximal Lie-nilpotent.

(iv) The Cartan subalgebras are exactly the centralizers of the radical complements.

(v) The Lie nilradical is exactly $rad(A) \oplus (Z(A) \cap T)$.

(vi) The number of isomorphic classes of maximal Lie nilpotent subalgebras is finite, and an upper bound is $B(dim_K(T))$.

(vii) The number of subalgebras in (iii) with respect to the nilradical and a fixed radical complement is finite and identically to the number in (ii).

(viii) $\mathcal{A}_M = \bigcup_{r \in rad(A)} (\mathcal{A}_{MT})^{1+r}$

(ix) Let M be a maximal Lie nilpotent subalgebra. The upper Lie central chain of M is $Z_n(M^\circ) = Z_n(rad(M)^\circ) \oplus VSEP(M)$ for all $n \in \mathbb{N}$. In particular, $cl(M^\circ) = cl(rad(M)^\circ) \leq cl(rad(M))$ is valid.

(x) If M is a maximal Lie nilpotent subalgebra, then $1 \leq cl(C_A(T)^\circ) \leq cl(M^\circ) \leq cl(nil(A^\circ)) = cl(rad(A)^\circ) \leq cl(rad(A))$ is valid.

Thus, the Fitting subalgebra possesses maximal and the Cartan subalgebras possess minimal Lie nilpotency class within all maximal Lie nilpotent subalgebras.

(xi) The attractor and repeller characteristics of maximal Lie nilpotent subalgebras within corollaries 6 and 7 are valid.◇

6.8 Open-ended questions and exercises

Open-ended questions 5 *(i) What is the exact number of isomorphism classes of maximal nilpotent Lie subalgebras in our context of solvable associative algebras? How many conjugacy classes do exist?*

(ii) If two maximal nilpotent Lie subalgebras are Lie isomorphic, are they then isomorphic as associative algebras, too?

(iii) What is the exact quantity of the sets within theorem 30?

(iv) Is it possible to describe the distribution of the Lie nilpotency classes of maximal Lie nilpotent subalgebras?

(v) What is the minimal resp. maximal dimension of all maximal Lie nilpotent subalgebras?

(vi) Is it possible to describe the distribution of the dimensions of all maximal Lie nilpotent subalgebras?

(vii) What is the exact attracting section of a maximal nilpotent Lie subalgebra?

(viii) Determine the iterative centralizers for the subalgebras of the nilradical powers and the descending and ascending central chain of the nilradical!

(ix) Is lemma 6 still true without assuming the solvability of the algebra?

(x) Is it possible to describe those subalgebras in a radical complement or in the nilradical which are fixed under the double-centralizer action?

(xi) Determine all maximal Lie nilpotent subalgebras of tensor products of associative (solvable) algebras.

(xii) Is the Chinese Remainder Theorem also valid for non-unital algebras?

(xiii) Determine those nilpotent Lie algebras L for which $cl(L)$ is exactly the maximum of all nilpotency classes of the elements $ad(l), l \in L$.

Excercise 97 *Determine for the radical $J(A)$ of the Solomon-Tits algebra, of the Solomon algebra in characteristic zero and of the algebras of upper and lower triangular matrices based on a field whether $cl(J(A)^\circ)$ is exactly the maximum of all nilpotency classes of the elements of $ad(j), j \in J(A)$.*

Excercise 98 *Let us focus on the topic within the previous exercise 97 for special group algebras: Let G be a cyclic p-group generated by z and K be a field of characteristic p. Then KG decomposes into $rad(KG)$ and $K1_G$. The radical is exactly $KG(1-z)$ (see e.g. [76]).*

Excercise 99 Let K be a field possessing at least three elements, A a finite-dimensional associative solvable unitary K-algebra possessing a separable factor algebra by its nilradical, T an algebra complement of $rad(A)$ in A and M a maximal Lie nilpotent subalgebra of A° such that $VSEP(M) \leq T$ is valid. Prove the following statements:

(i) $Z(A) \cap T \leq VSEP(M) \leq T$

(ii) $C_{rad(A)}(T) \leq C_{rad(A)}(VSEP(M)) \leq rad(A)$.

Illustrate both statements within a Hasse-diagram. What is the relevance of these statements?

Excercise 100 Formulate Eisenstein's criterion and give some examples for its usage.

Excercise 101 Prove part (iii) of remark 7 in details.

Excercise 102 Let A be an associative three dimensional algebra based on an infinite field generated by a nilpotent element of class 4. Investigate the subalgebras generated by two different elements of the set $\{x, x^2, x^3\}$. Which of these subalgebras possess finite many subalgebras?

Excercise 103 Within theorem 30 find a bijection between $\mathcal{A}_{DJ}$ and $\mathcal{A}_{DT}$!

Excercise 104 Based on the attractor and repeller properties find a connection to dynamical systems.

Excercise 105 Is it possible to generalize part (iii) of remark 7 from two to an arbitrary finite number of variables and an arbitrary field of characteristic p?

Excercise 106 Let p, q prime numbers. Prove that $\mathbb{Q}(\sqrt{p}, \sqrt{q})$ is simple and generated by $\sqrt{p} + \sqrt{q}$.

Excercise 107 In view of part (i) and (iii) of remark 7 analyze the following questions:

(i) For all $n \in \mathbb{N}$ the identity $B(n+1) = \sum_{k=1}^{n} \binom{n}{k} B(k)$ is valid.

(ii) For all $n \in \mathbb{N}$ the identity $B(n+1) > B(n)$ is valid. What is the meaning of this statement within remark 7?

(iii) Find all $n \in \mathbb{N}$ such that $B(n+1) > 2 \cdot B(n)$ is valid. What is the meaning of this statement within remark 7?

(iv) Generalize the first example in part (iii) to an arbitrary simple and finite-dimensional field extension of an arbitrary field!

Excercise 108 *Prove definition and remark 4 in details!*

Excercise 109 *Prove proposition 12 in details!*

Excercise 110 *Prove proposition 13 in details!*

Excercise 111 *Prove the Chinese Remainder Theorem for non-unital algebras for two comaximal ideals!*

Excercise 112 *Do a research in the literature for the opposite statement of the Chinese Remainder Theorem. Prove the opposite statement for two ideals within a finite-dimensional associative algebra.*

Excercise 113 *Prove $\mathbb{R}[t]/((t^2-1)\mathbb{R}[t]) \cong \mathbb{R}[t]/((t-1)\mathbb{R}[t]) \oplus \mathbb{R}[t]/((t+1)\mathbb{R}[t])$! Is it possible to generalize this exercise to an arbitrary polynomial in one variable over an arbitrary field? Transfer this exercise to $\mathbb{Z}$ and use the result on $\mathbb{Z}/(15 \cdot \mathbb{Z})$.*

Excercise 114 *Let A be an associative K-algebra and I, J K-ideals of A. Is $(I+J)/(I \cap J)$ isomorphic to $(I/(I \cap J)) \oplus (J/(I \cap J))$?*

Excercise 115 *Within $\mathbb{Z}$ focus on the ideals generated by 3 and by 6. Is the Chinese Remainder Theorem valid for these ideals? Find a pre-image for $(2+3\mathbb{Z}, 3+6\mathbb{Z})$ under χ!*

Excercise 116 *There are certain things whose number is unknown. If we count them by threes, we have two left over; by fives, we have three left over; and by sevens, two are left over. How many things are there?*

Excercise 117 *Within theorem 28 prove that $S \oplus K1_A$ is isomorphic to the algebra S^K which is the adjunction of an unit (see e.g. series 1 or [75]).*

Excercise 118 *Within example 5 try to find an associative algebra possessing the required characteristics.*

Excercise 119 *Which arguments used in remark 6 are valid in a more general context (e.g. for finite-dimensional division algebras)?*

Excercise 120 *Prove that a finite-dimensional field extension is simple if and only if only finite many intermediate fields exist. (Tip: see exercise 121)*

Excercise 121 Let $(K; K(a))$ be a finite-dimensional simple field extension and M an intermediate subfield of L containing K. Prove that M is determined by $\min_{a,M}$ and that $\min_{a,K}$ is divided by $\min_{a,M}$ in M. Is it possible to estimate the number of intermediate subfields by this approach?

Excercise 122 Formulate and prove the theorem of Engel for Lie algebras. In what way is the theorem relevant within this chapter?

Excercise 123 Prove corollary 3 in details!

Excercise 124 Prove part (ii) of lemma 11 in details.

Excercise 125 Prove corollary 6 in details.

Excercise 126 Prove example 5 in details.

Excercise 127 Does a natural number $n \in \mathbb{N}$ exists such that $B(n+1) = B(n) + 2^n - 1$ is valid?

Excercise 128 Prove corollary 7 in details!

Excercise 129 Formulate and prove the lemma of Goursat for algebras. By using the lemma of Goursat prove for an associative futile K-algebra A that $A \times K$ is futile, too. Use this result to prove that K^n is futile.

Excercise 130 Formulate and prove the lemma of Goursat for groups!

Excercise 131 Let A be an associative K-algebra, $x, y, l \in A$ and $r \in \mathbb{N}$. Prove the identity $(xy)(ad(l)^r) = \sum_{i=0}^{r} \binom{r}{k} x(ad(l)^k) y(ad(l)^{r-k})$.

Excercise 132 True or false: An associative algebra is futile if and only if its adjunction of an unit is futile.

Excercise 133 Prove remark 6 (if necessary by doing a research in the literature).

Excercise 134 Determine all maximal Lie nilpotent subalgebras of direct products of associative (solvable) algebras.

Excercise 135 Determine some maximal Lie nilpotent subalgebras of tensor products of associative (solvable) algebras.

Excercise 136 Analyze the determination of all non-unital subalgebras of K^n based on their orthogonal primitive idempotents. What is their number?

Excercise 137 *Determine the number of ideals and of the unital and non-unital subalgebras of K^n for $n \leq 10$. For this use and prove the formula $B(n+1) = \sum\limits_{k=0}^{n} \binom{n}{k} B(k)$.*

Excercise 138 *Are subalgebras of direct products exactly the direct products of subalgebras of the factors? What is the answer for groups?*

Excercise 139 *Determine all ideals of direct products of associative algebras and prove that these are exactly the direct products of ideals of the factors. (Tip: Focus on the projection of all factors.) Which theorem in this chapter is enhanced by this result? Is this result also true for groups and normal subgroups (instead of associative algebras and ideals)?*

Excercise 140 *Prove examples 6 and 7 in details. Is it possible to transfer the results to the upper triangular matrices? Within example 7 execute the iterative centralizer forming by using one of the 2^3 ideals of $D(3, K)$. What are the results for the other non-unital subalgebras of $D(3, K)$?*

Excercise 141 *Analyze exercise 140 for $K\Pi_3$!*

Excercise 142 *Analyze exercise 140 for D_3!*

Excercise 143 *Within theorem 30 determine the inverse function of the conjugation by $1 + r$. (Tip: What is the inverse of $1 + r$?)*

Excercise 144 *We assume the preconditions of theorem 30. Focus on the function from $\mathcal{A}_{MT}$ into $\mathcal{A}_{DT} \times \mathcal{A}_{DJ}$, $M \mapsto (VSEP(M); rad(M))$. Is this function really a function? Is the function injective, surjective or bijective? Is it possible to compare the order of the sets $\mathcal{A}_{MT}$ and $\mathcal{A}_{DT} \times \mathcal{A}_{DJ}$*

Excercise 145 *Let K be a field possessing at least three elements and A a finite-dimensional associative unitary solvable K-algebra possessing a separable factor algebra by its nilradical. Is it possible to transfer maximal Lie nilpotent subalgebras in a natural manner to ones of factor algebras, subalgebras, ideals, right ideals or left ideals of A?*

Excercise 146 *Let K be a field possessing at least three elements and A a finite-dimensional associative unitary solvable K-algebra possessing a separable factor algebra by its nilradical. Is there a connection between maximal Lie nilpotent subalgebras of A° and of $(A^K)^\circ$?*

Excercise 147 *Within lemma 10 prove part (ii) by using the argumentation of part (i).*

Excercise 148 *(zero-extension) Within exercise 61 try to determine all maximal nilpotent Lie subalgebras!*

Excercise 149 *(eAe) Within exercise 60 try to determine all maximal nilpotent Lie subalgebras!*

Excercise 150 *Determine examples of maximal Lie nilpotent subalgebras for $K\Pi_3$, D_3 and $\delta_{u.3}$ which are not related to the Lie nilradical and to the Cartan subalgebras!*

Excercise 151 *Let A be a finite-dimensional associative unital K-algebra. Is every maximal Lie subalgebra of $A°$ abelian?*

Excercise 152 *Within lemma 8 prove all used characteristics for the Jordan decomposition.*

Excercise 153 *Formulate the double-centralizer theorem for central division algebras. Is the theorem still valid if the roles of the simple and central-simples subalgebras are interchanged? Is the theorem still true if central-simple subalgebras are replaced by simple subalgebras?*

Excercise 154 *Prove exercise 8 on page 84 in [56]. In what way is this exercise relevant within this chapter?*

Excercise 155 *Prove remark 8 in details!*

Excercise 156 *Formulate and prove the following transformation formula: $C_T(J)^{1+r} = C_{T^{1+r}}(J^{1+r})$*

Excercise 157 *Do Lie nilpotent associative finite-dimensional unitary algebras exist such that their unit group is not nilpotent? What is the answer for the opposite implication?*

Excercise 158 *Are subalgebras of separable algebras separable, too?*

Excercise 159 *Is every finite-dimensional associative unitary algebra K-linear generated by their group of units? What is the answer for solvable or splitting solvable algebras? What is the connection to group algebras for these kind of algebra?*

Excercise 160 *Within theorem 24 analyze whether subalgebras of the form $N \oplus C$ such that the conditions $C_T(C_{rad(A)}(C)) = C$ and $C_{rad(A)}(C_T(N)) = N$ are valid are maximal Lie nilpotent.*

Excercise 161 *Find within the textbooks 'Basic Algebra I+II' of Nathan Jacobson all results and remarks concerning double-centralizers. What is there relevance for the density theorem of Jacobson?*

Excercise 162 Let K be a field of characteristic $p > 0$, $n \in \mathbb{N}$ and $f := t^n - 1 \in K[t]$. The factor algebra $K(t)/(f \cdot K(t))$ is semisimple if and only if n is no divisor of p. How many ideals and idempotents does the factor algebra $K(t)/(f \cdot K(t))$ possess? Bound the quantity of unital and non-unital subalgebras of $K(t)/(f \cdot K(t))$!

Excercise 163 Focus on the polynomial $f := t^5 - 1$ over $\mathbb{Q}$. How many ideals and idempotents does the factor algebra $K(t)/(f \cdot K(t))$ possess? Bound the quantity of unital and non-unital subalgebras of $K(t)/(f \cdot K(t))$!

Excercise 164 We assume the preconditions of theorem 30, and let M be an unital subalgebra of A containing the nilradical of A. Determine the centralizer of the radical complement of M in M! Compare it with the centralizer in A! Are these centralizers maximal Lie-nilpotent? What is the importance of these centralizers? Are the nilradicals of A and M connected?

Excercise 165 Analyze exercise 164 by using a subalgebra containing a radical complement (instead of containing the radical)!

Excercise 166 We assume the preconditions of theorem 30. Analyze for maximal nilpotent Lie subalgebras the usage of the centralizers in A instead of using the ones in T and $rad(A)$.

Excercise 167 Determine all maximal nilpotent subalgebras of $(\delta_{u,4})^\circ$ and the attraction sections of all unital subalgebras of the subalgebra of diagonal matrices.

Excercise 168 Prove proposition 11 in details!

Excercise 169 Prove remark 8 in details!

Chapter 7

A correspondence theorem between maximal nilpotent subgroups and Lie subalgebras

7.1 The correspondence theorem

The following lemma was proven already in series I and is needed for the next theorem in which maximal nilpotent Lie subalgebras are connected to maximal nilpotent subgroups:

Lemma 17 *Let K be a field possessing at least three elements and A a finite-dimensional associative unitary K-algebra possessing a separable factor algebra by its nilradical. A° is nilpotent if and only if $E(A)$ is nilpotent. If one of these statements is valid, then the associative algebra A is solvable and possesses (exactly one) central radical complement.*⋄

Based on lemmata 4 and 17 as well on remark 2 we can construct maximal nilpotent Lie subalgebras by using maximal nilpotent subgroups:

Theorem 31 *Let K be a field possessing at least three elements, A a finite-dimensional associative unitary solvable K-algebra possessing a separable factor algebra by its nilradical. If N is a maximal nilpotent subgroup of $E(A)$, then the following statements are valid:*

(i) $\langle N \rangle_K$ is an associative unital subalgebra of A.

(ii) $\langle N \rangle_K = \langle E(\langle N \rangle_K) \rangle_K$

(iii) $\langle N \rangle_K{}^\circ$ is a maximal nilpotent Lie subalgebra of A°.

(iv) $N = E(\langle N \rangle_K)$.

Proof. ad(i): Part (i) is a consequence of the fact that N is a subgroup of $E(A)$.

ad(ii): Part (ii) is deductable by remark 2.

ad(iii): By using lemma 4 we conclude that $\langle N \rangle_K{}^\circ$ is nilpotent. Lemma 17 lets us deduce that $E(\langle N \rangle_K)$ is nilpotent, too. N is contained in this group of units and is maximal nilpotent. Thus, the proof of part (iii) is finished.

ad(iv): Let L be a nilpotent Lie subalgebra of A° containing $\langle N \rangle_K$. We choose L maximal, and by using theorem 23 the Lie subalgebra L is an unital associative subalgebra. Lemma 17 implies that $E(L)$ is nilpotent, and by using remark 2 the identity $L = \langle E(L) \rangle_K$ is valid. We conclude $N = E(\langle N \rangle_K) \leq E(L)$, and by using the maximal nilpotency of L we derive $N = E(L)$. By this we deduce $\langle N \rangle_K = \langle E(L) \rangle_K = L$. Hence $\langle N \rangle_K$ is maximal Lie nilpotent.⋄

Definition 5 Let A be an associative unitary K-algebra. By $\mathcal{E}(\mathcal{A})_M$ we denote the set of all maximal nilpotent subgroups of $E(A)$.⋄

Within the next main theorem maximal Lie nilpotent subalgebras and maximal nilpotent subgroups are connected. By this and by the analysis of maximal Lie nilpotent subalgebras within the last chapter we are able to analysis maximal nilpotent subgroups and to transfer the results already proven for maximal nilpotent subalgebras to these subgroups. This analysis and transfer will be done within the next chapter. The theorem of Xiankun Du is used within the next main theorem connecting the class of nilpotency of maximal nilpotent subalgebras and subgroups. Cartan subalgebras and Carter subgroups as well as the nilradical and the Fitting subgroup are connected, too.

Main theorem 4 *Let K be a field possessing at least three elements, A a finite-dimensional associative unitary solvable K-algebra possessing a separable factor algebra by its nilradical. The maps*

$$t_M : \mathcal{A}_M \longrightarrow \mathcal{E}(\mathcal{A})_M, M \mapsto E(M) \text{ and } u_M : \mathcal{E}(\mathcal{A})_M \longrightarrow \mathcal{A}_M, U \mapsto \langle U \rangle_K$$

are bijections which are inverse to each other.

Let T be a radical complement of $rad(A)$ in A and M a maximal Lie nilpotent subalgebra of A° such that $VSEP(M) \leq T$ is true. The following statements are valid:

(i) $M = rad(M) \oplus VSEP(M)$ and $E(M) = (1+rad(M)) \times E(VSEP(M))$ are valid.

(ii) For all $n \in \mathbb{N}$ the identities $Z_n(M^\circ) = Z_n(rad(M)^\circ) \oplus VSEP(M)$ and $Z_n(E(M)) = (1+Z_n(rad(M)^\star)) \times E(VSEP(M)) = (1+Z_n(rad(M)^\circ)) \times E(VSEP(M))$ are valid.

(iii) $cl(M^\circ) = cl(rad(M)^\circ) = cl(rad(M)^\star) = cl(1+rad(M)) = cl(E(M)) \leq cl(rad(A))$

(iv) Cartan subalgebras and Carter subgroups are connected with respect to t_M and u_M.

(v) The nilradical and the Fitting subgroup are connected with respect to t_M and u_M.

Proof. By using theorem 31 we conclude that the map u_M is well-defined and that for every maximal nilpotent subgroup U of $E(A)$ the identity $Uu_M t_M = U$ is valid.
Let M be a maximal nilpotent Lie subalgebra of A°. Main theorem 3 implies that the Lie subalgebra M is an associative unital subalgebra, $M = rad(M) \oplus VSEP(M)$ is valid and $VSEP(M)$ is a separable commutative subalgebra of $Z(M)$. A consequence of this is the statement $E(M) = (1 + rad(M)) \times E(VSEP(M))$, and $E(VSEP(M))$ is a central normal subgroup of $E(M)$ (see e.g. [76]). In particular, part (i) is proven and $E(M)$ is nilpotent. By this and main theorem 1 about the theorem of Xiankun Du we conclude parts (ii) and (iii). Parts (iv) and (v) are consequences of the theorems 16, 14, 21 and 20.
We prove that $E(M)$ is maximal nilpotent. Let $E(M) \leq U \leq E(A)$ for a maximal nilpotent subgroup U of $E(A)$. By using remark 2 and Lemma 4 we derive that $M = \langle E(M) \rangle_K \leq \langle U \rangle_K \leq \langle E(A) \rangle_K = A$ is valid and that $\langle U \rangle_K$ is Lie nilpotent. The maximality of M implies $M = \langle U \rangle_K$, and we derive $E(M) = E(\langle U \rangle_K)$. Theorem 31 implies $U = E(\langle U \rangle_K)$, and we conclude $E(M) = U$. In addition, the map t_M is well-defined and by using remark 2 the identity $M = \langle E(M) \rangle_K$ is valid. This identity is equivalent to $Tt_M u_M = T$.⋄

We use this main theorem and focus on example 7. The group of units of the maximal Lie nilpotent subalgebras are examples of 5 maximal nilpotent subgroups. This example is analyzed further in chapter 8, but the reader can do this as an exercise within the next group of exercises, too.

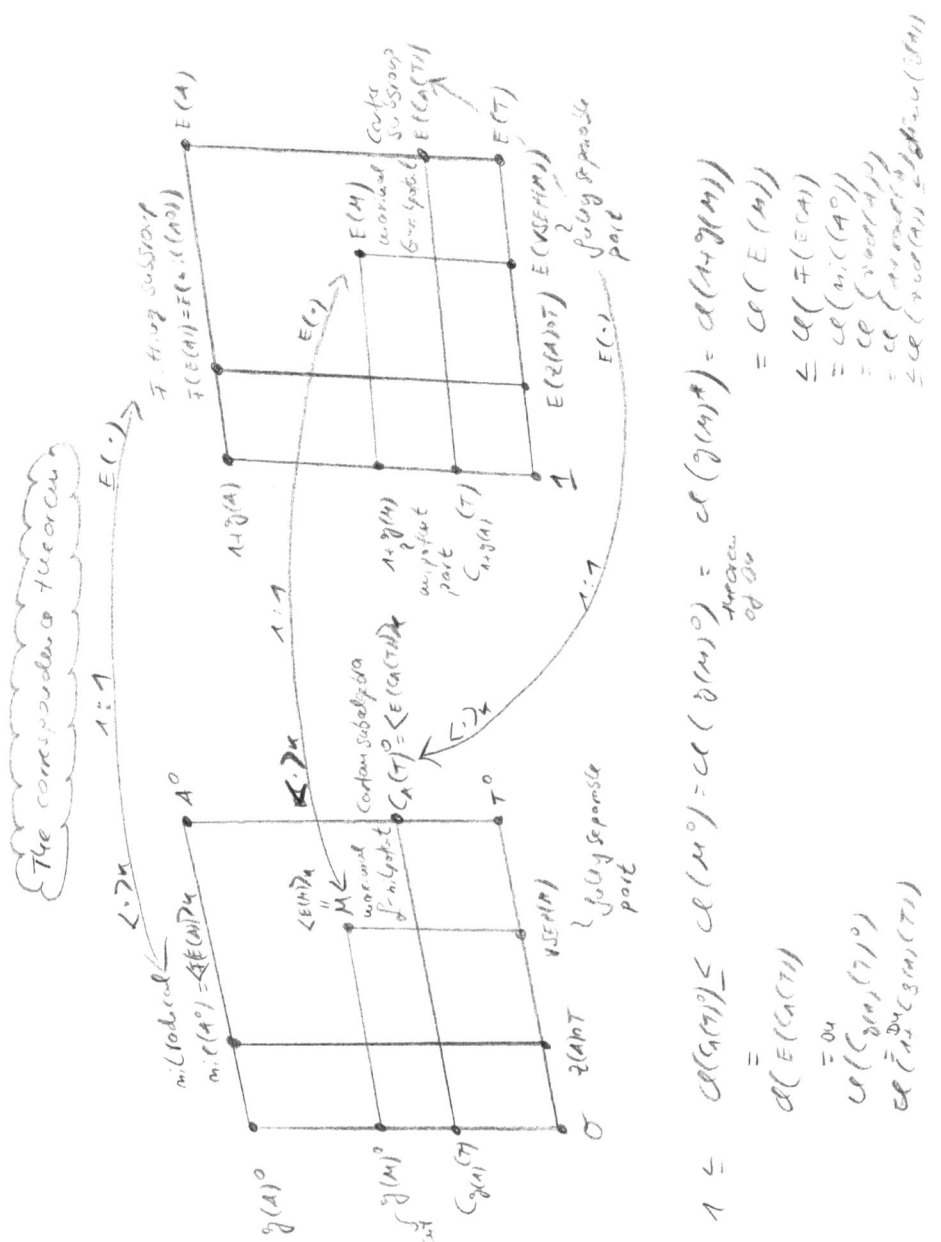

7.2 Open-ended questions and exercises

Open-ended questions 6 *(i) Is main theorem 4 also valid for other classes of algebras?*

(ii) Does a correspondence as described in theorem 4 exist for other properties of algebras like solvability?

Excercise 170 *Apply main theorem 4 to example 7!*

Excercise 171 *Apply main theorem 4 to example 6!*

Excercise 172 *Apply main theorem 4 to example 4!*

Excercise 173 *Apply main theorem 4 to example 5!*

Excercise 174 *(eAe) Apply main theorem 4 to exercise 60!*

Excercise 175 *(zero-extension) Apply main theorem 4 to exercise 61!*

Excercise 176 *Within main theorem 4 analyze the correspondence of ideals and normal subgroups!*

Excercise 177 *(correspondence for maximal abelian substructures) Let K be a field possessing at least three elements and A a finite-dimensional associative unitary solvable K-algebra possessing a separable factor algebra by its nilradical. The aim of this exercise is to determine the maximal commutative Lie subalgebras of A° and maximal abelian subgroups of $E(A)$. In addition, a correspondence between them is to be proven:*

(i) Every maximal Lie abelian subalgebra M of A° is an associative subalgebra of the form $M = rad(M) \oplus VSEP(M)$ such that $rad(M)$ is commutative.

(ii) If M is a maximal Lie abelian subalgebra of A°, then M is unital, and $E(M)$ is a maximal abelian subgroup of $E(A)$.

(iii) If U is a maximal abelian subgroup of $E(A)$, then $\langle U \rangle_K$ is a maximal abelian subalgebra of A°.

(iv) The functions $E(\cdot)$ and $\langle \cdot \rangle_K$ are inverse bijections between the sets of maximal Lie abelian subalgebras of A° and the maximal abelian subgroups of $E(A)$.

(v) An associative commutative subalgebra of the form $M = rad(M) \oplus VSEP(M)$ (such that $VSEP(M)$ is contained in a radical complement T) is maximal Lie abelian, if and only if $VSEP(M) = C_T(rad(M))$ is valid and $rad(M)$ is maximal commutative in $C_{rad(M)}(VSEP(M))$.

(vi) If N is a maximal commutative subalgebra of $rad(A)$, then $N \oplus C_T(N)$ is maximal Lie abelian.

(vii) Let I be a commutative subalgebra of $rad(A)$. Form $C_T(I)$ and afterwards $C_{rad(A)}(C_T(I))$. In this double-centralizer choose a maximal commutative subalgebra N. Then $N \oplus C_T(N)$ is maximal Lie abelian. All maximal Lie abelian subalgebras such that their fully separable part is contained in T are determinable by this process.

Is there a connection to the maximal commutative unital subalgebras of A?

Chapter 8

Maximal nilpotency in unit groups of solvable associative algebras

In this chapter we transfer the results of maximal nilpotent Lie subalgebras to maximal nilpotent subgroups by using the correspondence theorem proven within chapter 7. Some results are proven in details, others may be proven by the reader to experience the transfer principle based on the correspondence theorem.

8.1 A direct decomposition

Definition and remark 6 Let A be an associative unitary K-algebra and $u \in A$. u is called unipotent, if $u-1$ is nilpotent. Within the context of finite-dimensional associative unitary solvable K-algebras we have proven in series I that the nilradical is exactly the set of nilpotent elements. Under these assumptions u is unipotent if and only if $u \in 1 + rad(A)$ is valid. Unipotent elements are the key element for connecting the additive to the multiplicative Jordan decomposition. If $a = a_{nil} + a_{vsep}$ is a Jordan decomposition (see e.g. [75]) such that $a_{vsep} \in E(A)$ is true, then $a = a_{vsep} \cdot (1 + (a_{vsep})^{-1} a_{nil})$ is valid, a_{vsep} is fully separable and $1 + (a_{vsep})^{-1} a_{nil}$ is unipotent. Both elements commute with each other within this multiplicative decomposition. A structural connection of this decomposition within maximal nilpotent subgroups is a direct composition into two subgroups. This statement will be proven within the next theorem based on the correspondence main theorem 4. ⋄

Theorem 32 *Let K be a field possessing at least three elements, A a finite-dimensional associative unitary solvable K-algebra possessing a separable factor algebra by its nilradical and U a maximal nilpotent subgroup of $E(A)$.*

Exactly one maximal nilpotent Lie subalgebra M of $A°$ exists such that the following statements are valid:

(i) $U = E(M) = (1 + rad(M)) \times E(VSEP(M))$

(ii) $E(VSEP(M))$ is central in U.

(iii) $E(VSEP(M))$ is exactly the set of fully-separable invertible elements of M.

(iv) $1 + rad(M)$ is exactly the set of unipotent elements of M.

Proof. Based on main theorem 4 we know that exactly one maximal nilpotent Lie subalgebra of $A°$ exists such that $U = E(M)$ is valid. By using main theorem 3 we derive that M is an associative subalgebra, that $M = rad(M) \oplus VSEP(M)$ is valid and that $VSEP(M)$ is central in M. A semidirect decomposition of the algebra A leads to a semidirect decomposition of its group of units (see e.g. [76]). The fully separable elements of M are central in M, and thus the theorem is proven.⋄

If U is a maximal nilpotent subgroup within theorem 32, then we denote by $U = U_\mathcal{U} \times U_\mathcal{V}$ the multiplicative (Jordan) decomposition of U, and within this decomposition $U_\mathcal{U} = 1 + rad(M)$ is the unipotent and $U_\mathcal{V} = E(VSEP(M))$ the fully separable factor ($M := \langle U \rangle_K$ is the corresponding maximal Lie nilpotent subalgebra of $A°$.).

The following remark (which is based on remark 2) is needed to transfer results from Lie algebra to group theory and may be done by the reader as an exercise:

Remark 9 *Let K be a field possessing at least three elements, A a finite-dimensional associative unitary solvable K-algebra and T an unital subalgebra of A. The statements $T = \langle E(T) \rangle_K$ and $E(C_A(T)) = C_{E(A)}(E(T))$ are valid.*⋄

8.2 Manifold centralizers

We begin this section by proving that maximal nilpotent subgroups satisfy centralizer and double-centralizer properties. The following lemma is the pendant of lemma 9:

Lemma 18 *Let K be a field possessing at least three elements, A a finite-dimensional associative unitary solvable K-algebra possessing a separable factor algebra by its nilradical and U a maximal nilpotent subgroup of $E(A)$. The following statements are valid:*

(i) $C_{1+rad(A)}(U_\mathcal{V}) = U_\mathcal{U}$

(ii) $C_{E(T)}(U_\mathcal{U}) = U_\mathcal{V}$

(iii) $C_{E(T)}(C_{1+rad(A)}(U_\mathcal{V})) = U_\mathcal{V}$

(iv) $C_{1+rad(A)}(C_{E(T)}(U_\mathcal{V})) = U_\mathcal{U}$.

Proof. By using theorem 32 a maximal nilpotent Lie subalgebra M of A° exists such that $U = E(M) = U_\mathcal{U} \times U_\mathcal{V}$ and $M = rad(M) \oplus VSEP(M) = \langle E(M) \rangle_K$ are valid. Lemma 9 lets us derive the identities $C_{rad(A)}(VSEP(M)) = rad(M)$ and $C_T(rad(M)) = VSEP(M)$. $VSEP(M)$ is K-linear generated by its group of units (see remark 9), and thus we derive part (i) by using $C_{rad(A)}(VSEP(M)) = rad(M)$ and the shifting by the unit 1. If we apply the creation of the group of units to $C_T(rad(M)) = VSEP(M)$, then we derive again by using remark 9 (the second part within the remark) part (ii). Parts (iii) and (iv) are a direct consequence of parts (i) and (ii).⋄

The next lemma – which is the pendant of lemma 10 – shows us how to construct maximal nilpotent subgroups:

Lemma 19 *Let K be a field possessing at least three elements, A a finite-dimensional associative unitary solvable K-algebra possessing a separable factor algebra by its nilradical, T a radical complement of $rad(A)$ in A, C a subgroup of $E(T)$ and N a subgroup of $1+rad(A)$. The following statements are valid:*

(i) *If $C = C_{E(T)}(C_{1+rad(A)}(C))$ is valid, then $C \times C_{1+rad(A)}(C)$ is a maximal nilpotent subgroup of $E(A)$.*

(ii) *If $N = C_{1+rad(A)}(C_{E(T)}(N))$ is valid, then $C_{E(T)}(N) \times (N)$ is a maximal nilpotent subgroup of $E(A)$.*

Proof. The proof is a consequence of the analogue of this lemma – which is lemma 9 – and of theorem 32 and remark 9. Part (ii) is proven by part (i) as done in lemma 9. Thus we will only prove part (i) here. We have to derive that $C \times C_{1+rad(A)}(C)$ is a maximal nilpotent subgroup of $E(A)$. Based on theorem 32 it is sufficient to show that its K-linear span is a maximal nilpotent Lie subalgebra. This span is $\langle C \rangle_K \oplus C_{rad(A)}(\langle C \rangle_K)$. We apply lemma 9 on this K-linear span. Thus, we have to prove that $\langle C \rangle_K = C_T(C_{rad(A)}(\langle C \rangle_K))$ is valid. But this statement is a consequence of the assumption of (i) and the first part within remark 9 applied to T and to $E(T)$.⋄

The next theorem shows us that we can characterize maximal nilpotent subgroups by certain centralizer properties. By using lemma 19 and lemma 18 we derive directly:

Theorem 33 *Let K be a field possessing at least three elements, A a finite-dimensional associative unitary solvable K-algebra possessing a separable factor algebra by its nilradical, T a radical complement of $rad(A)$ in A and $M = U \times V$ a nilpotent subgroup of $E(A)$ such that $U \leq 1 + rad(A)$ and $V \leq E(VSEP(T))$ are valid. The following statements are equivalent:*

(i) M is maximal nilpotent.

(ii) $U = C_{1+rad(A)}(V)$ and $V = C_{E(T)}(U)$ are valid.

Add-on: If part (i) or part (ii) is valid, then $C_{E(T)}(C_{1+rad(A)}(U)) = U$ and $C_{1+rad(A)}(C_{E(T)}(V)) = V$ are true.⋄

The construction of subgroups possessing the double-centralizer property is proven within the next lemma which is the analogue of lemma 11:

Lemma 20 *Let K be a field, A a finite-dimensional associative K-algebra possessing a separable factor algebra by its nilradical, T a radical complement of $rad(A)$ in A, C a subgroup of $E(T)$ and N a subgroup of $1 + rad(A)$. The following statements are valid:*

(i) $C_{E(T)}(C_{1+rad(A)}(C_{E(T)}(C_{1+rad(A)}(C)))) = C_{E(T)}(C_{1+rad(A)}(C))$

(ii) $C_{1+rad(A)}(C_{E(T)}(C_{1+rad(A)}(C_{E(T)}(N)))) = C_{1+rad(A)}(C_{E(T)}(N))$

Proof. The proof can be done analogue to the one of lemma 11 and may be done by the reader within the exercises. Alternatively, the argumentation can be done based on the correspondence theorem.⋄

A direct consequence of the lemmata 18, 19 and 20 is the determination of all maximal nilpotent subgroups possessing a fully separable part contained in a fixed radical complement:

Theorem 34 *Let K be a field possessing at least three elements, A a finite-dimensional associative unitary solvable K-algebra possessing a separable factor algebra by its nilradical, T a radical complement of $rad(A)$ in A, C a subgroup of $E(T)$ and N a subgroup of $1+rad(A)$. The following statements are valid:*

(i) $C_{E(T)}(C_{1+rad(A)}(C)) \times C_{1+rad(A)}(C_{E(T)}(C_{1+rad(A)}(C)))$ is maximal nilpotent in $E(A)$.

(ii) $C_{1+rad(A)}(C_{E(T)}(N)) \times C_{E(T)}(C_{1+rad(A)}(C_{E(T)}(N)))$ is maximal nilpotent in $E(A)$.

Every maximal nilpotent subgroup of $E(A)$ such that the fully-separable is contained in $E(T)$ are creatable by this procedure.⋄

The theorem demonstrates us that we have to apply the double-centralizing once to all subgroups of the group of units of a special radical complement resp. of the normal subgroup $1 + rad(A)$. Afterwards lemma 19 is to be used to construct all maximal nilpotent subgroups based on the double centralized subgroups. All maximal nilpotent subgroups – possessing a fully separable part contained in a fixed radical complement – can be determined by this procedure which is the content of lemma 18. The same lemma lets us deduce that starting with subgroups in the radical or radical complement results at the end in the same maximal nilpotent subgroups.

The following attractor and repeller properties for the double-centralizing are valid which can be proven analogue to corollary 6 (and may be done by the reader within the exercises):

Corollary 11 *Let K be a field possessing at least three elements, A a finite-dimensional associative unitary solvable K-algebra possessing a separable factor algebra by its nilradical, T a radical complement of $rad(A)$ in A, C, D subgroups of $E(T)$ and N, M subgroups of $1 + rad(A)$. The following statements are valid:*

(i) $C \leq D \leq C_{E(T)}(C_{1+rad(A)}(C))$ results in
$C_{E(T)}(C_{1+rad(A)}(C)) = C_{E(T)}(C_{1+rad(A)}(D))$.

(ii) $N \leq M \leq C_{1+rad(A)}(C_{E(T)}(N))$ results in
$C_{1+rad(A)}(C_{E(T)}(N)) = C_{1+rad(A)}(C_{E(T)}(M))$.

(iii) $C_{E(T)}(C_{1+rad(A)}(C)) = C_{E(T)}(C_{1+rad(A)}(D))$ results in
$D \leq C_{E(T)}(C_{1+rad(A)}(C))$.

(iv) If $C_{1+rad(A)}(C_{E(T)}(N)) = C_{1+rad(A)}(C_{E(T)}(M))$ is valid,
then $M \leq C_{1+rad(A)}(C_{E(T)}(N))$ is true.

(v) If $C_{E(T)}(C_{1+rad(A)}(C)) = C_{E(T)}(C_{1+rad(A)}(D))$ is valid, then
$\langle C \cup D \rangle_\mathcal{G} \leq C_{E(T)}(C_{1+rad(A)}(C))$ is true.

(vi) If $C_{1+rad(A)}(C_{E(T)}(N)) = C_{1+rad(A)}(C_{E(T)}(M))$ is valid, then
$\langle N \cup M \rangle_\mathcal{G} \leq C_{1+rad(A)}(C_{E(T)}(N))$ is true.

(vii) If $C_{E(T)}(C_{1+rad(A)}(C)) < D$ is valid, then
$C_{E(T)}(C_{1+rad(A)}(C)) < C_{E(T)}(C_{1+rad(A)}(D))$ is true.

(viii) If $C_{1+rad(A)}(C_{E(T)}(N)) < M$ is valid, then
$C_{1+rad(A)}(C_{E(T)}(N)) < C_{1+rad(A)}(C_{E(T)}(M))$ is true.

(ix) If D is no subset of $C_{E(T)}(C_{1+rad(A)}(C))$, then $C_{E(T)}(C_{1+rad(A)}(D))$ is no subset of $C_{E(T)}(C_{1+rad(A)}(C))$, too.

(x) If N is no subset of $C_{1+rad(A)}(C_{E(T)}(N))$, then $C_{1+rad(A)}(C_{E(T)}(M))$ is no subset of $C_{1+rad(A)}(C_{E(T)}(N))$, too.◇

Definition 6 Let K be a field possessing at least three elements, A a finite-dimensional associative unitary solvable K-algebra possessing a separable factor algebra by its nilradical, T a radical complement, $U = C_{E(T)}C_{1+rad(A)}(U)$ a subgroup of $E(T)$ and $N = C_{1+rad(A)}C_{E(T)}(N)$ a subgroup of $1 + rad(A)$. The attraction section of U is the set of all subgroups V such that

$$C_{E(T)}C_{1+rad(A)}(U) = C_{E(T)}C_{1+rad(A)}(V)$$

is valid. The attraction section of N is set of subgroups M such that

$$C_{1+rad(A)}C_{E(T)}(N) = C_{1+rad(A)}C_{E(T)}(M)$$

is true.

Remark 10 Let K be a field possessing at least three elements, A a finite-dimensional associative unitary solvable K-algebra possessing a separable factor algebra by its nilradical, T a radical complement of $rad(A)$ in A, C, D subgroups of $E(T)$ and N, M subgroups of $1 + rad(A)$. Corollary 11 and lemma 20 let us construct different sequences of maximal nilpotent subgroups. In this remark we focus on sequences within the radical complement T. The reader may define analogue sequences within the exercises based on the nilradical. By using lemma 20 – beginning by using C – the subgroup $C_{E(T)}(C_{1+rad(A)}(C))$ is suitable to construct a maximal nilpotent subgroup:

$$C_{1+rad(A)}(C_{E(T)}(C_{1+rad(A)}(C))) \oplus C_{E(T)}(C_{1+rad(A)}(C)).$$

We define sequences of subgroups $(C_n)_{n \in \mathbb{N}}$ such that

$$C_{E(T)}(C_{1+rad(A)}(C_n)) = C_n$$

is valid, and thus

$$C_n \times C_{1+rad(A)}(C_{E(T)}(C_n))$$

are maximal nilpotent subgroups for all $n \in \mathbb{N}$.

Sequence 1:

We start by using $C \leq Z(E(A)) \cap T$. Thus, $C_{1+rad(A)}(C) = 1 + rad(A)$ is valid, and hence $C_1 := C_{E(T)}(C_{1+rad(A)}(C)) = C_{E(T)}(1 + rad(A)) = Z(E(A)) \cap T$ is true because of the solvability of A (T is commutative) and T being a radical complement. The maximal nilpotent subgroup associated to this double-centralizer is exactly the Fitting subgroup (see main theorem 4). Choose a subgroup D_1 containing $Z(E(A)) \cap T$ proper and minimal possessing this property, and construct $C_2 := C_{E(T)}(C_{1+rad(A)}(D_1))$. C_2 is containing C_1 proper (see corollary 11), and based on C_2 we are able to define another maximal nilpotent subgroup. Now choose D_2 containing C_2 proper and minimal possessing this property (if this is possible) and perform the same steps as defined previously for D_1. Because of the finite dimension of T we are reaching $E(T)$ after finite many steps using this construction: we focus on maximal Lie nilpotent subalgebras $\langle C_n \rangle_K$. These subalgebras are reaching T. Thus, applying the main theorem 4 their group of units are also reaching $E(T)$. These groups are exactly the subgroups C_n. The radical complement is associated to a Carter subgroup as described in main theorem 4.

Sequence 2:

We start by using the radical complement $C_1 := E(T)$. $E(T)$ is – as mentioned within the construction of sequence 1 – associated to a Carter subgroup. Choose a subgroup M_1 as small as possible of $E(T)$ possessing the property $C_{E(T)}(C_{1+rad(A)}(M_1)) = E(T)$. Choose D_1 as a subgroup of M_1 maximal possessing the property $C_2 := C_{E(T)}(C_{1+rad(A)}(D_1)) \neq E(T)$. This construction is repeated by using C_2 instead of $E(T)$. By this procedure the resulting subgroups C_n are getting smaller within each step. Hence, by using finite dimensionality and the argumentation used within sequence 1 the sequence of the subgroups C_n is reaching $Z(E(A)) \cap T$ after finite many steps. As described in sequence 1 the procedure is associated to the Fitting subgroup from now on and is constant.

Sequence 3:

Sequence 3 is a mixture of the sequences 1 and 2. Choose a subgroup C which is neither a Carter subgroup nor the Fitting subgroup (if such a subgroup is existing). Starting by C we use the procedure in sequence 1 reaching $E(T)$ after finite many steps which is associated to a Carter subgroup. From C down to $Z(E(A)) \cap T$ we define another sequence of subgroups in $E(T)$ analogue to the procedure of sequence 2. This sequence is constant of finite many steps, and the final subgroup in $E(T)$ is associated to the Fitting subgroup.

Carter subgroups and the Fitting subgroup are in view of these sequence extreme within all maximal nilpotent subgroups. For the Fitting subgroup its part in $E(T)$ is extremely small, but for the Carter subgroups extremely big. The reader may define analogue sequences within the nilradical. For them the situation is dual: for the Fitting subgroup the nilpotent part is extremely big, but for the Carter subgroups extremely small. The following corollary is related to this phenomena.⋄

By using main theorem 4 and corollary 7 we derive (The proof may be done by the reader within the exercises.):

Corollary 12 *Let K be a field possessing at least three elements, A a finite-dimensional associative unitary solvable K-algebra possessing a separable factor algebra by its nilradical, T a radical complement of $rad(A)$ in A, C a subgroup of $E(T)$ and N a subgroup of $1+rad(A)$. The following statements are valid:*

(i) $C_{E(T)}(C_{1+rad(A)}(C)) = Z(E(A)) \cap T$ *is valid if and only if C is a subgroup of $Z(E(A)) \cap T$.*
Exactly these subgroups are related to theorem 34 and are linked to the Fitting subgroup. (attracting section of the Fitting subgroup in $E(T)$)

(ii) $C_{1+rad(A)}(C_{E(T)}(N)) = C_{1+rad(A)}(E(T))$ *is valid if and only if N is a subgroup of $C_{1+rad(A)}(E(T))$.*
Exactly these subgroups are related to theorem 34 and are linked to the Carter subgroups. (attracting section of the Carter subgroups in $1+rad(A)$)

(iii) $C_{1+rad(A)}(C_{E(T)}(N)) = 1 + rad(A)$ *is valid if and only if $C_{E(T)}(N)$ is central in $E(A)$.*
Exactly these subgroups are related to theorem 34 and are linked to the Fitting subgroup. (attracting section of the Fitting subgroup in $1+rad(A)$)

(iv) $C_{E(T)}(C_{1+rad(A)}(C)) = E(T)$ *is valid if and only if $C_{1+rad(A)}(C) = C_{1+rad(A)}(E(T))$ is true.*
Exactly these subgroups are related to theorem 34 and are linked to the Carter subgroups. (attracting section of the Carter subgroups in $E(T)$)⋄

The next proposition clarifies the independence of the previous results from a special radical complement. This may be proven by the reader within the exercises based on the theorem of Wedderburn-Malcev:

Proposition 14 *Let K be a field possessing at least three elements, A a finite-dimensional associative unitary solvable K-algebra possessing a separable factor algebra by its nilradical and $T, \hat{T}$ radical complements of $rad(A)$ in A. If $\hat{U}$ is a maximal nilpotent subgroup of $E(A)$ such that $\hat{U}_V \in E(\hat{T})$ is valid, then an element $r \in rad(A)$ exists such that $(\hat{U}_V)^{1+r} \in E(T)$ is true.*

By changing the radical complements only isomorphic copies of maximal nilpotent subgroups are created.⋄

145

Finally, we bound the nilpotency class of maximal nilpotent subgroups, and again the Carter subgroups and the Fitting subgroups are extremal with respect to these bounds. The proof is based on main theorem 4 and on the pendent of this corollary which is corollary 3. In addition, the theorem of Xiankun Du is the key element for proving a connection to the Lie algebra side (Again this result may be proven by the reader within the exercises):

Corollary 5 *Let K be a field possessing at least three elements, A a finite-dimensional associative unitary solvable K-algebra possessing a separable factor algebra by its nilradical, T a radical complement of $rad(A)$ in A, U a maximal nilpotent subgroup of $E(A)$ and $M = \langle U \rangle_K$ the correspondent maximal nilpotent Lie subalgebra of A°. The following statements are valid:*

(i) *The ascending central chain of U is exactly $Z_n(U) = Z_n(U_\mathfrak{U}) \oplus U_\mathcal{V}$ for all $n \in \mathbb{N}$. In particular, $cl(U) = cl(U_\mathfrak{U}) = cl(M^\circ) = cl(rad(M)^\circ) \leq cl(rad(M))$ is valid.*

(ii) $1 \leq cl(C_A(T)^\circ) = cl(C_{E(A)}(E(T))) \leq cl(M^\circ) = cl(U) \leq cl(F(E(A)) = cl(1 + rad(A)) = cl(nil(A^\circ)) = cl(rad(A)^\circ) \leq cl(rad(A))$. ⋄

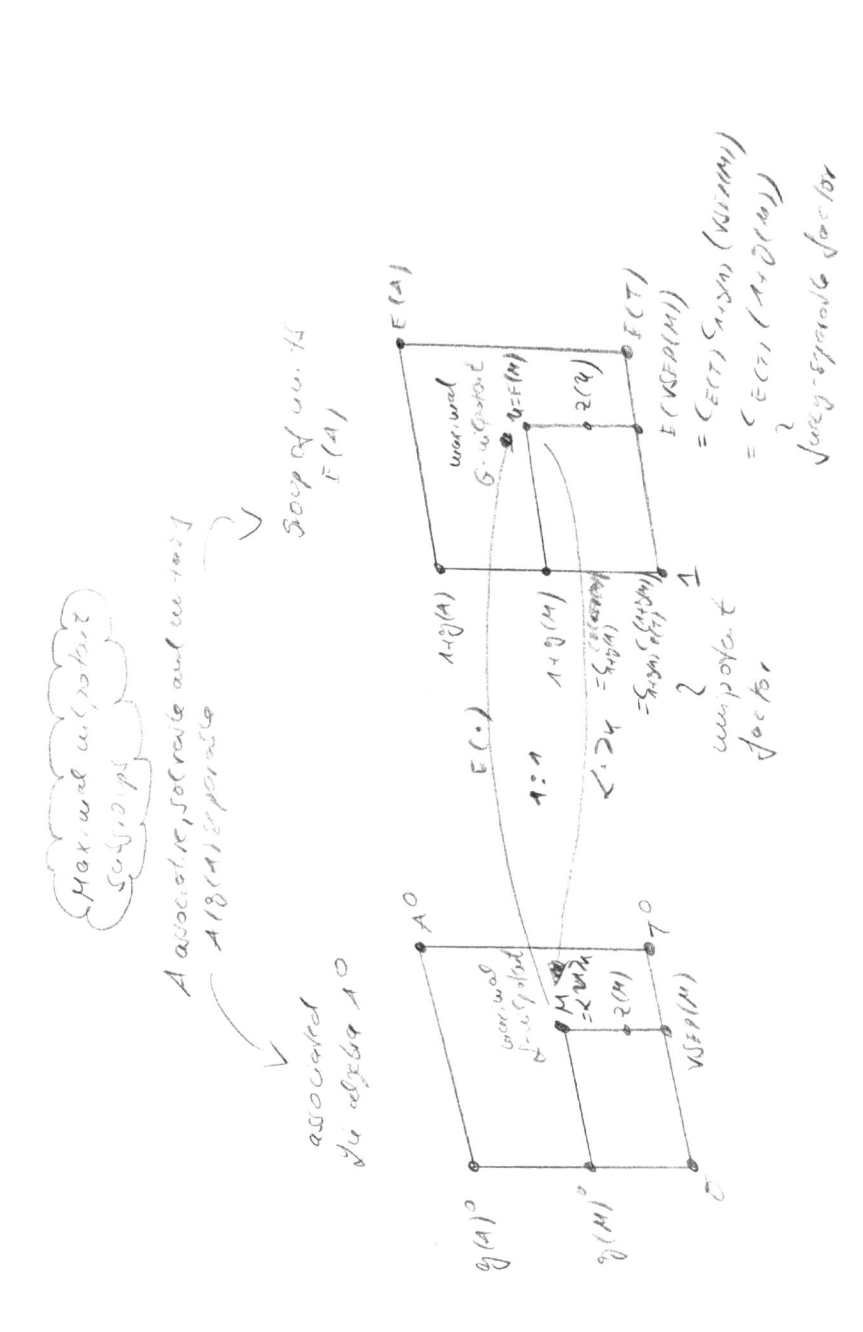

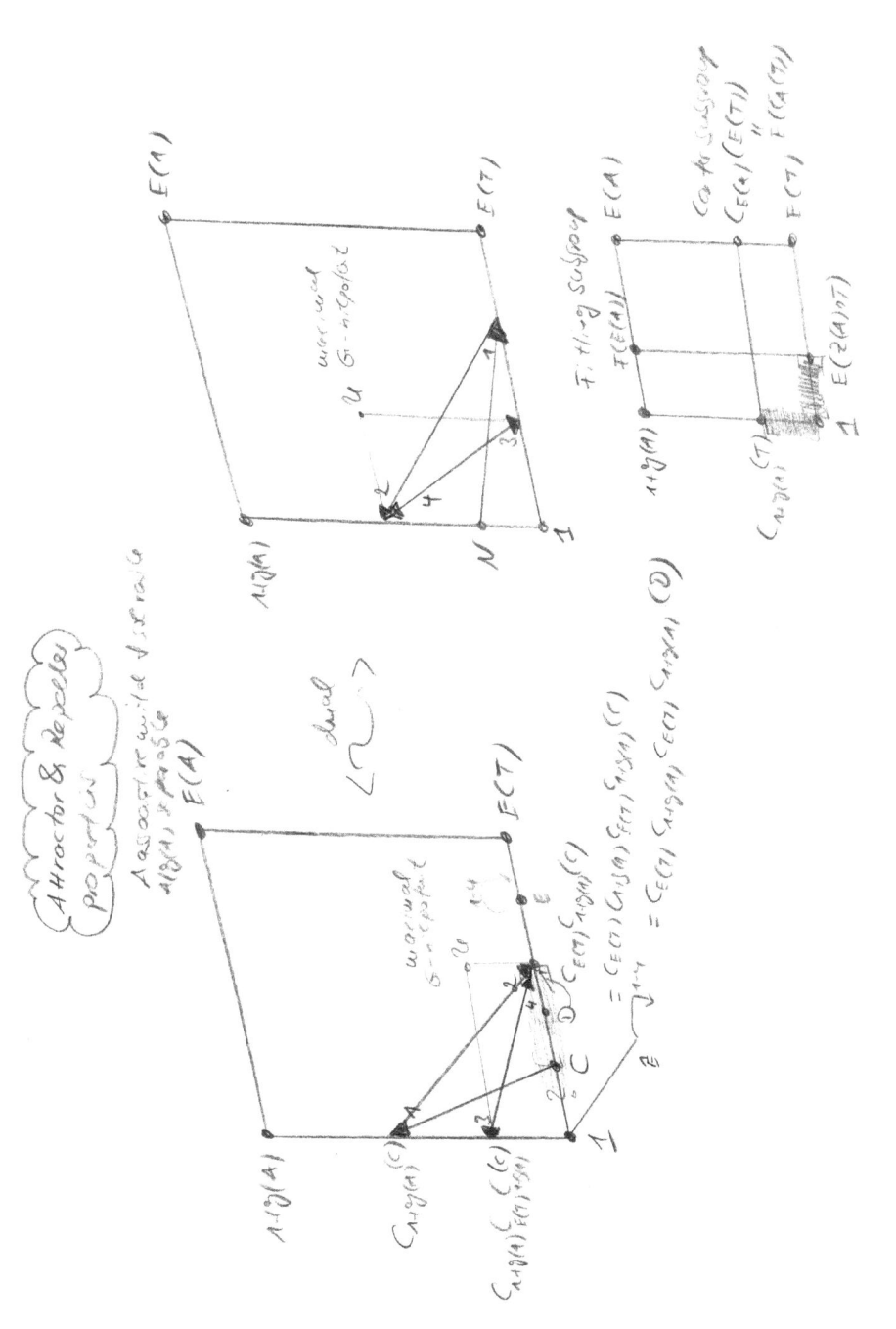

8.3 Finiteness of the number of isomorphism classes

Within this section we will prove that there are only finite many isomorphic classes of maximal nilpotent subgroups. Again, we will use the result for maximal Lie nilpotent subalgebras and main theorem 4.

Theorem 35 *Let K be a field possessing at least three elements, A a finite-dimensional associative solvable unitary K-algebra possessing a separable factor algebra by its nilradical, T a radical complement of $rad(A)$ in A, $\mathfrak{U}(E(T)) := \{C \mid C \leq E(T), C = C_{E(T)}(C_{1+rad(A)}(C))\}$ and $u_{E(T)} := |\mathfrak{U}(E(T))|$. The following statements are valid:*

(i) *There is exactly one (isomorphic class) Fitting subgroup in $E(A)$.*

(ii) *There is exactly one isomorphic class of Carter subgroups in $E(A)$.*

(iii) *There are only finite many isomorphic classes of maximal nilpotent subgroups of $E(A)$.*

(iv) *The number of isomorphism classes of maximal nilpotent subgroups of $E(A)$ can be estimated by the upper bound $u_{E(T)}$, and $u_{E(T)} = m_T \leq B(dim_K(T))$ is valid. In particular, the same upper bound is valid as for number of isomorphism classes of maximal nilpotent Lie subalgebras.*

(v) *If T is isomorphic to K^n for an element $n \in \mathbb{N}$, then $u_{E(T)} = m_T \leq B(n)$ is valid.*

Proof. ad(i): By definition at most one Fitting subgroups exists, and the existence is covered by theorem 21.

ad(ii): Part (ii) is a direct consequence of theorem 2.

ad(iii): This part is a consequence of part (iv).

ad(iv): We want to prove that the bijections within main theorem 4 are also bijections between $\mathfrak{U}(E(T))$ and $\mathfrak{M}(T)$. For this, $\mathfrak{M}(T) := \{C \mid C \leq T, C = C_T(C_{rad(A)}(C))\}$ is defined as used in theorem 29. Let $C \leq E(T)$ such that $C = C_{E(T)}(C_{1+rad(A)}(C))$ is valid. Hence, $D := \langle C \rangle_K$ is true by using remark 9, and the statement $D = C_T(C_{rad(A)}(D))$ is valid, too. If $D \leq T$ such that $D = C_T(C_{rad(A)}(D))$ is valid, then we deduce by using remark 9 the statement $E(D) = C_{E(T)}(C_{1+rad(A)}(E(D)))$. Let $C, U \leq E(T)$ such that the conditions $C = C_{E(T)}(C_{1+rad(A)}(C))$ and $U = C_{E(T)}(C_{1+rad(A)}(U))$ are valid. We assume $\langle C \rangle_K = \langle U \rangle_K$, and we have to prove the statement $C = U$. By using theorem 24 we derive that $\langle C \rangle_K \oplus C_{rad(A)}(\langle C \rangle_K)$ and $\langle U \rangle_K \oplus C_{rad(A)}(\langle U \rangle_K)$ are two identical maximal Lie nilpotent subalgebras.

Main theorem 4 lets us deduce that their group of units are identical, too. These subgroups are containing $C \times C_{1+rad(A)}(C)$ and $U \times C_{1+rad(A)}(U)$, and by using theorem 33 they are maximal nilpotent, too. Thus, we derive $U = E(\langle U \rangle_K) = E(\langle D \rangle_K) = D$. Now (ii) is proven using theorem 29.⋄

An open-ended question is whether the number of isomorphic classes for maximal nilpotent subgroups and Lie subalgebras is identical. Within the next remark we deal with this open-ended topic.

Remark 11 Let K be a field possessing at least three elements, A a finite-dimensional associative solvable unitary K-algebra possessing a separable factor algebra by its nilradical, T a radical complement of $rad(A)$ in A and M, N two maximal Lie-nilpotent subalgebras of $A°$. Based on the correspondence theorem 4 the Lie subalgebras M, N are unital associative subalgebras of A and their group of units $E(M), E(N)$ are maximal nilpotent subgroups of $E(A)$ such that $M = \langle E(M) \rangle_K$ and $N = \langle E(N) \rangle_K$. It is straightforward to prove that $E(M)$ and $E(N)$ are isomorphic as groups and M, N as Lie algebras if M and N are isomorphic as associative algebras. We will prove that M, N are isomorphic as associative algebras if $E(M), E(N)$ are isomorphic as groups. A consequence is that the number isomorphism classes of maximal Lie nilpotent subalgebras of $A°$ is not greater than the number of isomorphism classes of maximal nilpotent subgroups of $E(A)$. Still open is whether M, N are isomorphic as associative algebras if M, N are isomorphic as Lie algebras. The author conjectures that this statement is wrong. In case of the field $\mathbb{R}$ the group of units is the Lie group related to the Lie algebra. There are examples of not-isomorphic Lie groups possessing isomorphic Lie algebras.

Let $E(M)$ and $E(N)$ isomorphic as groups based on a group isomorphism α. Then $K(E(M))$ and $K(E(N))$ are isomorphic as associative algebras based on an isomorphism $\hat{\alpha}$ which is α restricted to $E(M)$. In addition, the map $\beta : K(E(M)) \longrightarrow \langle E(M) \rangle_K$ which is the identity map restricted on $E(M)$ is an associative epimorphism. Analogously, the map $\gamma : K(E(M)) \longrightarrow \langle E(M) \rangle_K$ which is the identity map restricted on $E(M)$ is an associative epimorphism. It is straightforward to prove that the map $\chi : K(E(M))/(ker(\alpha)) \longrightarrow \langle K(E(N))/(ker(\beta)), a + (ker(\alpha)) \mapsto a\hat{\alpha} + (ker(\beta))$ is an isomorphism. ⋄

8.4 Cardinalities

Within this section we analyze cardinalities of sets of subgroups which are relevant to determine maximal nilpotent subgroups within this chapter. In addition, we will connect these sets to the corresponding ones related to maximal nilpotent Lie subalgebras. The following definitions are a starting point of our analysis:

Definition and remark 7 Let K be a field, A a finite-dimensional associative unitary K-algebra possessing a separable factor algebra by its nilradical and T be a radical complement of $rad(A)$ in A. We define:

- $\mathcal{G}_{DT} := \{C \mid C \leq E(T), C_{E(T)}(C_{1+rad(A)}(C)) = C\}$
- $\mathcal{G}_{DJ} := \{C \mid C \leq 1 + rad(A), C_{1+rad(A)}(C_{E(T)}(C)) = C\}$ and
- $\mathcal{G}_{UT} := \{U \mid U \leq E(A), U = X \times Y, X \leq 1+rad(A), Y \leq E(T), C_{E(T)}(X) = Y, C_{1+rad(A)}(Y) = X\}$.

In addition, let $\mathcal{G}_U$ be the set of all maximal nilpotent subgroups. We have proven that we can use the first two sets $\mathcal{G}_{DT}$ and $\mathcal{G}_{DJ}$ to define elements within the third set $\mathcal{G}$. The next theorem shows us that this construction is complete. For this, we will use that all separable radical complements are conjugated under $1_A + rad(A)$ by the theorem of Wedderburn-Malcev. $\diamond$

The next theorem can be deduced by theorem 30 and its analogue to theorem 35. Within this analogue we have already proven that the sets $\mathcal{G}_{DT}$ and $\mathcal{M}_{DT}$ are of the same order based on the functions 'build the group of units' and 'generate the K-space'. This leads to the connection between group and Lie theory. The other statements are a consequence of this or can be deduced by the stated bijections (as done in theorem 30). All of these may be proven by the reader within the exercises.

Theorem 36 *Let K be a field possessing at least three elements, A a finite-dimensional associative solvable unitary K-algebra possessing a separable factor algebra by its nilradical, T, S radical complements of $rad(A)$ in A and $r \in rad(A)$ such that $S = T^{1+r}$ is true. The following statements are valid:*

(i) $\mathcal{G}_{UT}$ is the set of all maximal nilpotent subgroups U of $E(A)$ such that their fully separable part is contained in $E(T)$.

(ii) $\mathcal{G}_{UT}$ and $\mathcal{G}_{US}$ are finite sets of the same order. In details, the function $\mathcal{G}_{UT} \longrightarrow \mathcal{G}_{US}, M \mapsto M^{1+r}$ is a bijection.

(iii) $\mathcal{G}_{UT}$ and $\mathcal{A}_{MT}$ are finite sets of the same order based on the functions 'build the group of units' and 'generate the K-space'.

(iv) $\mathcal{G}_{DJ}$ and $\mathcal{G}_{UT}$ are finite sets of the same cardinality. In details, the function $\mathcal{G}_{DJ} \longrightarrow \mathcal{G}_{MT}, N \mapsto C_{E(T)}(N) \times N$ is a bijection possessing the inverse function $M \mapsto 1 + rad(M)$.

(v) $\mathcal{G}_{DJ}$ and $\mathcal{A}_{DJ}$ are sets of the same cardinality.

(vi) $\mathcal{G}_{DT}$ and $\mathcal{G}_{UT}$ are finite sets of the same order. In details, the function $\mathcal{G}_{DT} \longrightarrow \mathcal{G}_{UT}, C \mapsto C_{1+rad(A)}(C) \times C$ is a bijection possessing the inverse map $M \mapsto VSEP(M)$.

(vii) $\mathcal{G}_{DT}$ and $\mathcal{A}_{DT}$ are sets of the same cardinality.

(viii) $\mathcal{G}_{UT}$ and $\mathcal{G}_{DS}$ are finite sets of the same cardinality. In details, the function $\mathcal{G}_{DT} \longrightarrow \mathcal{G}_{DS}, C \mapsto C^{1+r}$ is a bijection.

(ix) $\mathcal{G}_{DS}$ and $\mathcal{A}_{DS}$ are sets of the same cardinality.

(x) The number of elements of $\mathcal{G}_{UT}$ can be estimated by the upper bound $B(dim_K(T))$.

(xi) $\mathcal{G}_U = \bigcup\limits_{r \in rad(A)} (\mathcal{G}_{UT})^{1+r}$ ⋄

Remark 12 We remark the following concerning theorem 36:

(i) If the nilradical is infinite, then $\mathcal{G}_U = \bigcup\limits_{r \in rad(A)} (\mathcal{G}_{UT})^{1+r}$ needs not to be finite. We have proven that each set of the union is finite and of the same cardinality. If K is finite, then we can estimate the order of $\mathcal{G}_U$ by the upper bound $|rad(A)| \cdot |\mathcal{G}_{UT}|$, and for this number $|rad(A)| \cdot B(dim_K(T))$ is an upper bound.

(ii) The intersection of the sets $\mathcal{G}_{UT}$ – such that T varies within all radical complements – are not known to the author. One known aspect is the following one: : If $r \in rad(A)$ and U is a maximal nilpotent subgroup such that $U \in \mathcal{G}_{UT} \cap \mathcal{G}_{UT}^{1+r}$ is valid, then the condition $r \in C_{1+rad(A)}(VSEP(U)) = U_\mathcal{U}$ is true. From this we derive $U = U^{1+r}$. If we want to construct potential new maximal subgroups by passing from U to a conjugate of U, then the conjugator must be chosen outside of $U_\mathcal{U}$.⋄

The proof of part (i) and (ii) may be done by the reader as exercises.⋄

Thickness of the unions of isomorphism classes

$\Lambda + g(A)$

T ng $subgrp$
$T(E(A))$

$N + D_N \times CE_D$ (N)   married subgrps

G_3g
subgrps
with double
central or
property

$E(A)$

$F(M)$

$E(C(T))$ Cart Subgrp

$E(T)$

G_0T — subgrps with double or no torsion property

$1,1 \to G_0T$

Δ, T

$\{$ number of isomorphism classes $\leq 8(\dim(T))$

$|G_0g| = |G_0T| = |G_0T|$
"
$|M_0g| = |M_0T| = |Aut|$

8.5 Examples

Within the following examples let K be a field possessin at least three elements, A a finite-dimensional associative solvable K-algebra possessing a separable factor algebra by its nilradical, T, S radical complements of $rad(A)$ in A and $r \in rad(A)$ mit $S = T^{1+r}$.

Example 9 *(maximal subgroup)* Let C be a maximal subgroup of $E(T)$ and N be a maximal subgroup of $1 + rad(A)$. The statements

$$C \leq C_{E(T)}(C_{1+rad(A)}(C)) \leq E(T)$$

and

$$N \leq C_{1+rad(A)}(C_{E(T)}(N)) \leq 1 + rad(A)$$

are valid. By using the maximality of the subgroups we derive that C and N are possessing the double-centralizer property or that their double-centralizers are $E(T)$ resp. $1 + rad(A)$. Within the first case the direct product $C \times C_{1+rad(A)}(C)$ resp. $N \times C_{E(T)}(N)$ is maximal nilpotent by using main theorem 4. In the other case the double-centralizing leads to a Carter subgroup resp. to the Fitting subgroup (again by using main theorem 4).⋄

Example 10 *(minimal non-nilpotent)* Let $E(A)$ be minimal non-nilpotent. We prove that the Fitting subgroup and the Carter subgroups are the only possible maximal nilpotent subgroups. If U is a maximal nilpotent subgroup of $E(A)$, then we focus on the subgroup $B := (1+rad(A)) \times U_\mathcal{V}$. If $B = E(A)$ is valid, then $U_\mathcal{V}$ is the group of units of a radical complement. In this case we obtain a Carter subgroup (see main theorem 4). In the other case B is nilpotent containing U. Thus, U and B are identical and we deduce $1 + rad(A) = U_\mathcal{U}$. In this case we obtain a Fitting subgroup (see main theorem 4).⋄

Example 11 *(triangular matrices I)* We focus on the subalgebra of lower triangular matrices $\delta_{u,3}$ and analyze the powers of the nilradical. For this, we use lemma 10 and corollary 7. The nilradical yields by double-centralizing to the Lie nilradical and the zero-space (which is $rad(A)^3$) to a Cartan subalgebra.

Now let us focus on the ideal $rad(A)^2$ which can be represented by the matrices of the form $\begin{pmatrix} 0 & 0 & 0 \\ 0 & 0 & 0 \\ a & 0 & 0 \end{pmatrix}$. $C_{D(3,K)}(rad(A)^2)$ is represented by the matrices of the form $\begin{pmatrix} x & 0 & 0 \\ 0 & y & 0 \\ 0 & 0 & x \end{pmatrix}$. In addition, $C_{rad(A)}(C_{D(3,K)}(rad(A)^2))$

is exactly $rad(A)^2$. Thus, $rad(A)^2 \oplus C_{D(3,K)}(rad(A)^2)$ is another maximal Lie nilpotent subalgebra of $\delta_{u,3}°$. (The calculations based on matrices are straightforward and may be done by the reader as exercises.)

The group of units of this maximal nilpotent Lie subalgebra is a maximal nilpotent subgroup. By using the determinant function the upper triangular matrices of the form $\begin{pmatrix} x & 0 & 0 \\ 0 & y & 0 \\ a & 0 & x \end{pmatrix}$ such that $x^2 y \neq 0$ – and hence $x \neq 0 \neq y$ – is valid represents this group of units. For this argumentation we use the statement that the group of units of an unital subalgebra of a finite-dimensional associative unitary algebra is exactly the intersection of this subalgebra with the group of units of the underlying algebra (see e.g. [77]). Thus we can use the determinant to describe the maximal nilpotent subgroup. The next example is the dual version of this example.◇

Example 12 *(triangular matrices II)* We focus again on the subalgebra of lower triangular matrices $\delta_{u,3}$ and analyze all unital subalgebras of $D(3, K)$. By using lemma 10, corollary 7, lemma 14 and theorem 29 we can bound the number of isomorphic classes of maximal Lie nilpotent subalgebras. There are 5 unital subalgebras T of $D(3, K)$. For each such subalgebra we calculate the double-centralizing and determine a maximal Lie nilpotent subalgebra. By this, we determine 5 maximal Lie nilpotent subalgebras, and these are the potential isomorphic classes of maximal Lie nilpotent subalgebras. (The calculations based on matrices are straightforward and may be done by the reader as exercises.) Let e_1, e_2, e_3 be the orthogonal primitive idempotents of $D(3, K)$.

$D(3, K)$ is the unique 3-dimensional unital subalgebra of $D(3, K)$. Double-centralizing yields to a Cartan subalgebra:

$$\begin{pmatrix} 0 & 0 & 0 \\ 0 & 0 & 0 \\ 0 & 0 & 0 \end{pmatrix} \oplus \begin{pmatrix} a & 0 & 0 \\ 0 & b & 0 \\ 0 & 0 & c \end{pmatrix}.$$

The other extreme is given by the K-subspace generated by the unit (which is $e_1 + e_2 + e_3$). The double-centralizing yields to the nilradical:

$$\begin{pmatrix} 0 & 0 & 0 \\ a & 0 & 0 \\ c & b & 0 \end{pmatrix} \oplus \begin{pmatrix} d & 0 & 0 \\ 0 & d & 0 \\ 0 & 0 & d \end{pmatrix}.$$

Now we have to focus on the 2-dimensional unital subalgebras of $D(3, K)$ which are $T_1 := \langle e_1 + e_2, e_3 \rangle_K$, $T_2 := \langle e_1 + e_3, e_2 \rangle_K$ and $T_3 := \langle e_1, e_2 + e_3 \rangle_K$. It is straightforward to calculate that for all $i \in \underline{3}$ the condition $C_{D(3,K)}(C_{rad(\delta_{u,3})}(T_i)) = T_i$ is valid. The double-centralizing is stable from the first step. Thus we derive the following maximal Lie nilpotent subalgebras represented by matrices:

$$\begin{pmatrix} 0 & 0 & 0 \\ a & 0 & 0 \\ 0 & 0 & 0 \end{pmatrix} \oplus \begin{pmatrix} b & 0 & 0 \\ 0 & b & 0 \\ 0 & 0 & c \end{pmatrix} \text{ (for } T_1\text{)},$$

$$\begin{pmatrix} 0 & 0 & 0 \\ 0 & 0 & 0 \\ a & 0 & 0 \end{pmatrix} \oplus \begin{pmatrix} b & 0 & 0 \\ 0 & c & 0 \\ 0 & 0 & b \end{pmatrix} \text{ (for } T_2\text{) and}$$

$$\begin{pmatrix} 0 & 0 & 0 \\ 0 & 0 & 0 \\ 0 & a & 0 \end{pmatrix} \oplus \begin{pmatrix} b & 0 & 0 \\ 0 & c & 0 \\ 0 & 0 & c \end{pmatrix} \text{ (for } T_3\text{).}\diamond$$

A subalgebra which is not fixed by the double-centralizing is e.g. an ideal of the radical complement $D(3, K)$ because ideals are not unital. This fact (for ideals) is true in a more general context, and the reader may calculate the double-centralizing for all ideals of $\delta_{u,3}$ within the exercises.

The corresponding groups of units of these 5 maximal Lie nilpotent subalgebras are maximal nilpotent subgroups, and by theorem 35 these are the potential isomorphic classes of maximal nilpotent subgroups. As described in example 12 be can use the determinant function to calculate the group of units. For all the same relevant condition is $b, c \neq 0.\diamond$

8.6 Summary

In this section we summarize the results related to maximal nilpotent subgroups:

Main theorem 5 *Let K be a field possessing at least three elements, A a finite-dimensional associative unitary solvable K-algebra possessing a separable factor algebra by its nilradical and T a radical complement. The following statements are valid by summarizing the results of the previous sections within this chapter:*

(i) *Every maximal nilpotent subgroup is the group of units of exactly one maximal Lie nilpotent subalgebra. In particular, the cardinality of maximal nilpotent subalgebras and subgroups is identical.*

(ii) *Every maximal nilpotent subgroup U of $E(A)$ possesses a decomposition $U = U_\mathcal{U} \times U_\mathcal{V}$ such that $U_\mathcal{U}$ is the set of unipotent element of U, $U_\mathcal{V}$ is the set of fully separable elements of U and $U_\mathcal{V}$ is contained in $Z(U)$. Up to conjugation by an element $1 + r, r \in rad(A)$ the subgroup $U_\mathcal{V}$ is contained in $E(T)$.*

(iii) If $U = X \times Y$ is a nilpotent subgroup of $E(A)$ such that $X \leq 1+rad(A)$ and $Y \leq E(T) \cap Z(U)$ are valid, then U is maximal nilpotent if and only if

$$X = C_{1+rad(A)}(Y) \text{ and } Y = C_{E(T)}(X)$$

are valid.

(iv) All maximal nilpotent subgroups such that their fully separable part is contained in $E(T)$ are constructible by double-centralizing of subgroups C of $E(T)$ resp. N of $1 + rad(A)$:
$C_{E(T)}(C_{1+rad(A)}(C)) \times C_{1+rad(A)}(C_{E(T)}(C_{1+rad(A)}(C)))$ and
$C_{1+rad(A)}(C_{E(T)}(N)) \times C_{E(T)}(C_{1+rad(A)}(C_{E(t)}(N)))$ are maximal nilpotent.

(v) All Carter subgroups are exactly the centralizers of the group of units of the radical complements. These are exactly the group of units of the Cartan subalgebras.

(vi) The Fitting subgroup is exactly $(1 + rad(A)) \times (E(Z(A) \cap T))$. Its the group of units of the Lie nilradical.

(vii) The number of isomorphic classes of maximal nilpotent subgroups of $E(A)$ is finite and can be estimated by the upper bound $B(dim_K(T))$.

(viii) The number of subgroups in (iii) with respect to the nilradical and the radical complement is finite and identical, and this number is identical to the number of subgroups in (ii).

(ix) $\mathcal{G}_U = \bigcup_{r \in rad(A)} (\mathcal{G}_{UT})^{1+r}$

(x) Let U be a maximal nilpotent subgroup of $E(A)$ and $M := \langle U \rangle_K$. The upper Lie central chain of U is exactly $Z_n(U) = Z_n(U_\mathcal{U}) \times U_\mathcal{V}$ for all $n \in \mathbb{N}$. In particular, by using the theorem of Xiankun Du we derive $cl(U) = cl(U_\mathcal{U}) = cl(M°) \leq cl(rad(M))$.

(xi) Let U be a maximal nilpotent subgroup of $E(A)$ and $M := \langle U \rangle_K$. The statement $1 \leq cl(C_{E(A)}(E(T))) \leq cl(U) = cl(M°) \leq cl(F(E(A))) = cl(1 + rad(A)) = cl(nil(A°)) = cl(rad(A)°) \leq cl(rad(A))$ is valid.

(xii) The attractor and repeller characteristics of the corollaries 11 and 12 are valid.⋄

8.7 Open-ended questions and exercises

Open-ended questions 7 *(i) Is the number of isomorphic classes of maximal nilpotent subgroups and Lie subalgebras identically in our context of solvable associative algebras? In other words, are two maximal nilpotent Lie subalgebras are Lie isomorphic if and only if their correspondent maximal nilpotent subgroups are isomorphic as groups?*

(ii) What is the exact number of isomorphism classes of maximal nilpotent subgroups in our context of solvable associative algebras? How many conjugacy classes are existing?

(iii) What is the exact number of sets within theorem 36?

(iv) What is the exact attracting section of a maximal nilpotent subgroup?

(v) Determine the iterative centralizers for the subgroups of the shifted nilradical powers by 1 and the descending and ascending central chain of the shifted nilradical by 1!

Excercise 178 *(correspondence of attraction section) Let K be a field possessing at least three elements, A a finite-dimensional associative unitary solvable K-algebra possessing a separable factor algebra by its nilradical, $C = C_T C_{rad(A)}(C)$ an unital subalgebra of T, $J = C_{rad(A)} C_T(J)$ a subalgebra of $rad(A)$, $U = C_{E(T)} C_{1+rad(A)}(U)$ a subgroup of $E(T)$ and $N = C_{1+rad(A)} C_{E(T)}(N)$ a subgroup of $1 + rad(A)$. The aim of this exercise is to prove the correspondence of the attractor sections of these subgroups and unital subalgebras. The attraction section of C is the set of all unital subalgebras D such that $C_T C_{rad(A)}(C) = C_T C_{rad(A)}(D)$ is valid. The attraction section of U is the set of all subgroups V such that $C_{E(T)} C_{1+rad(A)}(U) = C_{E(T)} C_{1+rad(A)}(V)$ is valid. Prove that a bijective correspondence between the attraction sections of C and $E(C)$ resp. of U and $\langle U \rangle_K$ exists. Define the attraction sections for J and $1 + J$ resp. N and $\langle N \rangle_K$ and determine a bijective correspondence between them.*
In addition, prove that for two subgroups within an attraction section their intersection need not belong to the same attraction section (Tip: example 8).

Excercise 179 *Prove remark 11 in details!*

Excercise 180 *We focus on the examples 7 and 12. Draw a picture for the correspondence established in exercise 178.*

Excercise 181 *We focus again on example 7. In this exercise the attraction section is to be calculated including all non-unital subalgebras. Within the example we have proven that there are exactly 5 unital subalgebras which are* stable under the double-centralizing process. Thus, their attraction section based on all unital subalgebras consists of a single element: the unital subalgebra itself. Prove – in more general context – that the attraction section including all non-unital subalgebras is exactly the set of non-unital subalgebras S such that $S \oplus K1_A$ is exactly the unital subalgebra. Draw a picture of the attraction section including all non-unital subalgebras and compare it with the picture stated in remark 7.

Excercise 182 Let K be a field possessing at least three elements and A a finite-dimensional associative unitary solvable K-algebra possessing a separable factor algebra by its nilradical. Analyze whether the following statement is true: A° is minimal non-nilpotent if and only if $E(A)$ is minimal non-nilpotent.

Excercise 183 Which properties of the additive Jordan decomposition (see e.g. in [75]) are valid for multiplicative one, too (e.g. uniqueness, existence for separable elements, representation by polynomials)

Excercise 184 Derive the multiplicative Jordan decomposition by the additive one and vice versa. What assumptions are necessary?

Excercise 185 Do a research in the literature concerning the theorem of Engel for groups, formulate and prove the theorem. Compare the proof with the corresponding one for Lie algebras.

Excercise 186 Prove corollary 5 in details.

Excercise 187 Prove lemma 20 in details.

Excercise 188 Prove corollary 11 in details.

Excercise 189 Prove the statements in example 10 concerning nilpotency.

Excercise 190 Prove corollary 12 in details.

Excercise 191 Let A be an associative K-algebra, $x, y, l \in A$ and $r \in \mathbb{N}$. The identity $(xy)(ad(l)^r) = \sum_{i=0}^{r} \binom{r}{k} x(ad(l)^k) y(ad(l)^{r-k})$ is valid. Does a pendant exist for groups?

Excercise 192 Describe maximal nilpotent subgroups of unit groups of direct products of associative unitary algebras.

Excercise 193 Try to determine a maximal nilpotent subgroup of unit groups of tensor products of associative solvable unitary algebras possessing a separable factor algebra by their nilradicals.

Excercise 194 *Prove examples 11 and 12 in details. Transfer the results to the algebra of upper triangular matrices. Determine within example 12 the iterative centralizers starting by using one the 2^3 ideals of $D(3, K)$. How can we describe this iterative process for all other non-unital subalgebras of $D(3, K)$?*

Excercise 195 *Analyze exercise 194 for $K\Pi_3$ instead of $\delta_{u,3}$!*

Excercise 196 *Analyze exercise 194 for D_3 instead of $\delta_{u,3}$!*

Excercise 197 *Determine within theorem 36 the inverse map for the conjugation by $1 + r$? (Tip: Determine the inverse of $1 + r$.)*

Excercise 198 *We assume the preconditions of theorem 36. Focus on the function from $\mathcal{G}_{UT}$ into $\mathcal{G}_{DT} \times \mathcal{G}_{DJ}$ defined by $U \mapsto (X; Y)$. Prove that its really a function and determine whether the function is injective, surjective or bijective. Are there some connections concerning the order of $\mathcal{G}_{UT}$ and $\mathcal{G}_{DT} \times \mathcal{G}_{DJ}$?*

Excercise 199 *Is maximal nilpotency compatible with the adjunction of an unit concerning Lie algebras and unit groups?*

Excercise 200 *Within lemma 19 prove part (ii) with a similar argument used for part (i).*

Excercise 201 *(zero-extension) Within exercise 61 try to determine all maximal nilpotent subgroups!*

Excercise 202 *(eAe) Within exercise 60 try to determine all maximal nilpotent subgroups!*

Excercise 203 *Determine examples of maximal nilpotent subgroups of $E(K\Pi_3)$, $E(D_3)$ and $E(\delta_{u.3})$ which are different from the Fitting subgroup and from the Carter subgroups!*

Excercise 204 *Is every maximal nilpotent subgroup of the unit group of an unitary associative solvable K-algebra abelian?*

Excercise 205 *Prove remark 12 in details!*

Excercise 206 *Transfer the following transformation rule to unit groups and prove it: $C_T(J)^{1+r} = C_{T^{1+r}}(J^{1+r})$*

Excercise 207 *Are there examples of Lie nilpotent associative algebras possessing non-nilpotent group of units? Are there examples for the opposite implication?*

Excercise 208 *Transfer exercise 164 to the group of units and prove the corresponding exercise.*

Excercise 209 *Analyze exercise 208 by using a subalgebra containing a radical complement (instead of containing the radical)!*

Excercise 210 *We assume the preconditions of theorem 36. Analyze for maximal nilpotent subgroups the usage of the centralizers in $E(A)$ instead of using the ones in $E(T)$ and $1 + rad(A)$?*

Excercise 211 *Prove main theorem 5 in details! Analyze the double-centralizer condition by starting with a normal subgroup of $E(A)$ contained in $E(T)$ resp. an ideal of A contained in T!*

Excercise 212 *Prove theorem 36 in details.*

Excercise 213 *Prove proposition 14 in details.*

Excercise 214 *Prove remark 9 in details.*

Excercise 215 *Is within the multiplicative and additive Jordan decomposition the identity $(a_{vsep})^{-1} = (a^{-1})_{vsep}$ valid? Determine and compare the additive and multiplicative Jordan decomposition of a^{-1} and of ka, $k \in K$.*

Excercise 216 *Prove the following statements in details: For this argumentation we use the fact that the group of units of an unital subalgebra of a finite-dimensional associative unitary algebra is exactly the intersection of this subalgebra with the group of units of the underlying algebra (see e.g. [77]). What is the relevance of this result within this chapter?*

Excercise 217 *Let K be a field and A a finite-dimensional associative unitary K-algebra. For each $a \in A$ let $a\rho$ be the function from A in A sending x to xa. Prove that the function ρ sending a to $a\rho$ is an algebra monomorphism from A into $End_K(A)$. Use exercise 216 to deduce that an element $a \in A$ is invertible if and only if $det(a\rho)$ is not zero. In addition, prove that if an element $a \in A$ is nilpotent, then $tr(a\rho) = 0$ is valid. (det is the determinant function and tr the trace function of the endomorphism of a K-space.)*

Excercise 218 *Let G be a group, K a field and ρ as defined in exercise 217. If $G := S_3 = \{1, (12), (123), (132), (13), (23)\}$, then use the determinant function to decide for $K := GF(2)$, $K := GF(3)$ and $K := \mathbb{C}$ which of the following elements are invertible in KG: $g \in G$, $1 + (12)$, $1 + (123)$ and $1 + (12) + (123)$. Is it possible to find a non-nilpotent element $x \in KG$ such that $tr(x\rho) = 0$?*

Excercise 219 Let G be a group, K a field and ρ as defined in exercise 217. If $G := V_4 = \{1, (12), (34), (12)(34)\}$, then use the determinant function to decide for $K := GF(2)$, $K := GF(5)$ and $K := \mathbb{C}$ which of the following elements are invertible in KG: $g \in G$, $1+(12)$, $1+(123)$ and $1+(12)+(123)$. Is it possible to find a non-nilpotent element $x \in KG$ such that $tr(x\rho) = 0$?

Excercise 220 What is the effect by using the left multiplication within the last three exercises?

Excercise 221 Transfer exercise 99 to the group of units.

Excercise 222 What are the extremal properties of Cartan subalgebras, Carter subgroups, the Lie nilradical and the Fitting subgroups among all maximal nilpotent subgroups resp. Lie subalgebras?

Chapter 9

Fischer subgroups, nilpotent projectors and injectors

9.1 Fischer subgroups

Definition and remark 8 A nilpotent subgroup F of a group G is called Fischer subgroup, if every nilpotent subgroup of G normalized by F is contained in F:

$$\forall U \leq G : U \text{ nilpotent} \wedge F \leq N_G(U) \longrightarrow U \leq F.$$

A Fischer subgroup is maximal nilpotent. Within finite solvable groups Bernd Fischer has proven that Fischer subgroups exist and form exactly one conjugacy class of subgroups. We want to determine the Fischer subgroups for unit groups of – not necessary finite – solvable associative algebras.⋄

Theorem 37 *Let K be a field possessing at least three elements and A a finite-dimensional associative unitary solvable K-algebra possessing a separable factor algebra by its nilradical. The Fitting subgroup is the only Fischer subgroup of $E(A)$.*

Proof. We prove at first that $F(E(A))$ is a Fischer subgroup of $E(A)$. $F(E(A))$ is a nilpotent normal subgroup of $E(A)$. Let U be a nilpotent subgroup of $E(A)$ such that $F(E(A)) \leq N_{E(A)}(U)$ is valid. By definition U is a normal subgroup of $N_{E(A)}(U)$, and we assume that U is nilpotent. The normal subgroup $F(E(A))$ is contained in $N_{E(A)}(U)$, too, and it is nilpotent. We conclude by using a theorem of Fitting that $U \cdot F(E(A))$ is a nilpotent normal subgroup of $N_{E(A)}(U)$. The Fitting subgroup is maximal nilpotent in $E(A)$ by theorem 21, and thus we derive $F(E(A)) = U \cdot F(E(A))$. Therefor $U \leq F(E(A))$ is valid, and $F(E(A))$ is a Fischer subgroup.

We have to prove that every Fischer subgroup is identical to $F(E(A))$. Let

F be a Fischer subgroup of $E(A)$. The normalizer of $F(E(A))$ in $E(A)$ is $E(A)$ because $F(E(A))$ is a normal subgroup of $E(A)$. Hence F is contained in this normalizer. By the definition of Fischer subgroups we derive that the nilpotent subgroup $F(E(A))$ is contained in F. F is nilpotent and contains $F(E(A))$. $F(E(A))$ is a maximal nilpotent subgroup by theorem 21, and thus $F = F(E(A))$ is valid.⋄

9.2 The pendant for Lie algebras: Fischer subalgebras

Definition and remark 9 Let L be a K-Lie algebra. A nilpotent subalgebra B of L is called Fischer subalgebra if B contains every nilpotent subalgebra normalized by B:

$$\forall T \leq L : \text{L nilpotent} \wedge B \leq N_L(T) \rightarrow T \leq B.$$

If M is a nilpotent subalgebra containing B, then $B \leq M \leq N_L(M)$ is valid. We conclude $B = M$, and thus B is maximal nilpotent. We analyze and determine the Fischer subalgebras of Lie algebras associated to solvable associative algebras.⋄

Theorem 38 *Let K be a field possessing at least three elements, A a finite-dimensional associative unitary solvable K-algebra possessing a separable factor algebra by its nilradical. The Lie nilradical is the only Fischer subalgebra of A°. In particular, the unit group of the Fischer subalgebra of A° is the Fischer subgroup of $E(A)$ and the K-space generated by the Fischer subgroup of $E(A)$ is the Fischer subalgebra of A°.*

Proof. We prove at first that the Lie nilradical is a Fischer subalgebra of A°. The Lie nilradical is a nilpotent ideal of A°, and by using theorem 20 it is a maximal nilpotent subalgebra of A°. Let T be a nilpotent subalgebra of A° normalized by the Lie nilradical $nil(A^\circ)$. $nil(A^\circ)$ and T are nilpotent ideals of the normalizer of T in A°, and hence by a theorem of Hans Fitting their sum is nilpotent, too. This sum contains $nil(A^\circ)$ and is nilpotent. $nil(A^\circ)$ is a maximal nilpotent subalgebra (see theorem 20), and we conclude that the sum is exactly the Lie nilradical, and T is contained in $nil(A^\circ)$.
We have to prove that every Fischer subalgebra B of A° is identical to $nil(A^\circ)$. $nil(A^\circ)$ is an ideal of A°, and hence its normalizer is A. A contains B, and we conclude $nil(A^\circ) \leq B$ by the definition of the Fischer subalgebra. By the maximal nilpotency of $nil(A^\circ)$ we derive $B = nil(A^\circ)$.⋄

Remark 13 In view of theorems 38 and 37 we conclude the following corollary for a field K possessing at least three elements and a finite-dimensional associative unitary solvable K-algebra A possessing a separable factor algebra by its nilradical:

(i) If B is a Fischer subalgebra of A°, then $E(B)$ is a Fischer subgroup of $E(A)$.

(ii) If F is a Fischer subgroup of $E(A)$, then $\langle F \rangle_K$ is a Fischer subalgebra of A°.

A proof for this statement without using theorems 38 and 37 is not known by the author. An open-ended question is whether this corollary is true in a more general context of Lie algebras associated to associative algebras.⋄

9.3 Nilpotent projectors

Definition and remark 10 Let G be a group and P a subgroup of G. P is called nilpotent projector, if $(PN)/N$ is maximal nilpotent in G/N for every normal subgroup N of G. In particular, P is a maximal nilpotent subgroup of G. Within finite solvable groups the Carter subgroups are exactly the nilpotent projectors.⋄

Theorem 39 *Let K be a field possessing at least three elements, A a finite-dimensional associative unitary solvable K-algebra possessing a separable factor algebra by its nilradical. Every nilpotent projector is a Carter subgroup.*

Proof. Let P be a nilpotent projector of $E(A)$. By definition and remark 10 the subgroup P is a maximal nilpotent subgroup of $E(A)$. Using the nilpotent normal subgroup $1 + rad(A)$ and the definition of the nilpotent projector we derive that $(P(1 + rad(A)))/(1 + rad(A))$ is maximal nilpotent in $E(A)/(1 + rad(A))$. This factor group is abelian and hence $(P(1 + rad(A)))/(1 + rad(A)) = E(A)/(1 + rad(A))$ is valid. We conclude $E(A) = (1 + rad(A)) \cdot P$. By using the main theorem 4 we derive $A = \langle E(A) \rangle_K = rad(A) + \langle P \rangle_K$, and $\langle P \rangle_K$ is maximal Lie-nilpotent and an associative subalgebra of the form $\langle P \rangle_K = rad(\langle P \rangle_K) \oplus VSEP(\langle P \rangle_K)$. We deduce $A = rad(A) \oplus VSEP(\langle P \rangle_K)$, and hence $VSEP(\langle P \rangle_K)$ is a radical complement. Again, by using the main theorem 4 we derive that $\langle P \rangle_K$ is a Cartan subalgebra, and we conclude by the same theorem that $P = E(\langle P \rangle_K)$ is a Carter subgroup of $E(A)$. ⋄

Example 13 It is not known by the author whether the Carter subgroups possess the characteristics of a nilpotent projector for infinite fields under the assumptions of theorem 39. The following examples provided by Derek Holt demonstrates that this is not true for an arbitrary infinite solvable group (see [86]):

This is not true for infinite insoluble groups.

Let K be the direct product of countably infinitely many copies $\langle a_i, b_i \rangle$ ($i \in \mathbb{N}$) of $S_3 = \langle a, b \mid a^2 = b^3 = (ab)^2 = 1 \rangle$, and let G be the semidirect product of K with a group $\langle t \rangle$ of order 2 that acts on each copy of S_3 by $a_i^t = a_i b_i$, $b_i^t = b_i^{-1}$. (So t is acting in the same way on $\langle a_i, b_i \rangle$ as conjugation by $a_i b_i^{-1}$.)

Then the subgroup $A = \langle a_i \mid i \in \mathbb{N} \rangle$ is nilpotent (it's an elementary abelian 2-group) and self-normalizing in G. But $N = \langle b_i \mid i \in \mathbb{N} \rangle$ (which is an elementary abelian 3-group) is normal in G, G/N is abelian, and $(AN)/N$ is normal of index 2 in G/N.⋄

9.4 The pendant for Lie algebras: nilpotent Lie projectors

Definition and remark 11 Let L be a K-Lie algebra and P a subalgebra of L. P is called nilpotent projector if for every ideal I of L the subalgebra $(P+I)/I$ is maximal nilpotent in L/I. In particular, P is maximal nilpotent in L.⋄

Theorem 40 *Let K be a field possessing at least three elements, A a finite-dimensional associative unitary solvable K-algebra possessing a separable factor algebra by its nilradical. The nilpotent projectors of A° are exactly the Cartan subalgebras of A°. Their group of units are the Carter subgroups which are the possible candidates for the nilpotent projectors of $E(A)$.*

Proof. We prove at first that Cartan subalgebras are the only candidates for nilpotent projectors of A°. Let P be a nilpotent projector of A°. By using definition and remark 11 the projector P is maximal nilpotent in A°, and we conclude by the main theorem 4 that it is an associative subalgebra of the form $P = rad(P) \oplus VSEP(P)$, such that $VSEP(P)$ is central in P and $rad(P) \leq rad(A)$ is valid. If we apply the projector identity to $I := rad(A)$, then we deduce that $(P + rad(A))/rad(A)$ is maximal Lie-nilpotent in $A/rad(A)$. This factor algebra is commutative by the solvability of A, and hence we deduce $(P + rad(A))/rad(A) = A/rad(A)$ and $A = P + rad(A)$. Therefor $A = rad(P) \oplus VSEP(P) + rad(A)$ is valid, and we derive $A = VSEP(P) \oplus rad(A)$. Hence $VSEP(P)$ is a radical complement of $rad(A)$ in A. Again by using main theorem 4 the Cartan subalgebras are the only maximal Lie-nilpotent subalgebras with this characteristic.

In series I we have proven within corollary 20 that for a sufficiently large field K the Cartan subalgebras are compatible with factor algebras: if C is a Cartan subalgebra of A° and I an ideal of A°, then $(C + I)/I$ is a Cartan subalgebra of A°/I. In particular, this subalgebra is maximal nilpotent within this factor algebra. This compatibility of Cartan subalgebras with factor algebras is by using [6] true in a more general context without the assumption for the field size. Theorem 39 finishes the proof.⋄

9.5 Nilpotent injectors

Definition and remark 12 Let G be a group and I a subgroup of G. I is called nilpotent injector, if for every normal subgroup N of G the intersection $I \cap N$ is maximal nilpotent in G/N. In particular, (for $N := G$) I is a maximal nilpotent subgroup of G.
Within finite solvable groups the nilpotent injectors are exactly the maximal nilpotent subgroups containing the Fitting subgroup, and they are all conjugated.⋄

Remark 14 *(Dedekind identity)* Let G be a group and A, B, C subgroups of G such that $C \leq A$ is true. The Dedekind identity is valid:

$$A \cap (B \cdot C) = (A \cap B) \cdot C.\diamond$$

Theorem 41 *Let K be a field possessing at least three elements, A a finite-dimensional associative unitary solvable K-algebra possessing a separable factor algebra by its nilradical. The Fitting subgroup of $E(A)$ is the only nilpotent injector.*

Add-on: For every normal subgroup N of $E(A)$ the identity $F(E(A)) \cap N = F(N)$ is valid, and every subgroup N containing $F(N)$ is a normal subgroup of $E(A)$.

Proof. $F(E(A))$ is a nilpotent normal subgroup of $E(A)$. Let I be a nilpotent injector of $E(A)$. $I \cap F(E(A))$ is maximal nilpotent in $F(E(A))$, and hence this intersection is $F(E(A))$. Thus, $F(E(A))$ is contained in the nilpotent subgroup I, and by theorem 16 the subgroups I and $F(E(A))$ must be identical.

Let N be a normal subgroup of $E(A)$. $F(E(A)) \cap N$ is a nilpotent normal subgroup of N. Let U be a nilpotent subgroup situated between $F(E(A)) \cap N$ and N. We prove that U is a normal subgroup of $E(A)$. By using this statement the proof of the theorem is finished. Let $g \in E(A)$ and $u \in U$. The identity $u^g = [g, u^{-1}] \cdot u$ is valid. The derivation of $E(A)$ is – by using theorem 16 – contained in $1 + rad(A)$, and this nilpotent normal subgroup is contained – again by using theorem 16 – in $F(E(A))$. Hence $u^g \in F(E(A)) \cdot U$ is valid. In addition, $u^g \in N^g = N$ is true because N is a normal subgroup. We deduce $u^g \in (F(E(A)) \cdot U) \cap N$. By using $U \leq N$ and the Dedekind identity of remark 14 we conclude $u^g \in (F(E(A)) \cap N) \cdot U \leq U$, and hence U is a normal subgroup of $E(A)$.⋄

9.6 The pendant for Lie algebras: nilpotent Lie injectors

Definition and remark 13 Let L be a K-Lie algebra and I a subalgebra of L. I is called nilpotent Lie injector if for every ideal J of L the intersection $I \cap J$ is maximal nilpotent in J. In particular, for $J := L$ the subalgebra I is maximal nilpotent in L.⋄

Theorem 42 *Let K be a field possessing at least three elements, A a finite-dimensional associative unitary solvable K-algebra possessing a separable factor algebra by its nilradical. The nilradical of A° is the only nilpotent Lie injector of A°. Its group of units is the only nilpotent injector of $E(A)$ and coincide with the Fitting subgroup of $E(A)$.*

Add-on: For every ideal J of A° the subalgebra $nil(A^\circ) \cap J$ is exactly $nil(J)$, and every ideal I of J containing $nil(A^\circ) \cap J = nil(J)$ is an ideal of A°.

Proof. We prove at first that the nilradical of A° is a nilpotent Lie injector of A°. Let I be a nilpotent Lie injector of A°. Then the intersection $nil(A^\circ) \cap I$ is maximal nilpotent in $nil(A^\circ)$ which is nilpotent, too. Hence the nilradical is contained in the maximal nilpotent Lie subalgebra I. By using theorem 20 both substructures must be identical.

Let J be an ideal of A°. We want to prove that $nil(A^\circ) \cap J$ is maximal nilpotent in J. Let $nil(A^\circ) \cap J \leq T \leq J$ such that T is a nilpotent subalgebra of J. We conclude $T \circ A \leq A \circ A \leq rad(A) \leq nil(A^\circ)$ because A is solvable. In addition, $T \circ A \leq J \circ A \leq J$ is valid because J is an ideal. Therefor $T \circ A$ is contained in $nil(A^\circ) \cap J$ which is contained in T by our assumption. We have proven that T is a nilpotent ideal of A° contained in J. We derive $T \leq nil(A^\circ) \cap J$, and thus $nil(A^\circ) \cap J$ is maximal nilpotent in J and identical to $nil(J)$. The proof is finished by using theorem 41.⋄

(Extremal normal nilpotent substructors)

Finite subgroup(s) $\xrightarrow{E(\cdot)}$ Finite subgroup(s)
$\wr$ $1:1$ $\wr$
Finitely subgroup   Vi-reduced
$\wr$ $\wr$
Ni potent projectif(s) $\xrightarrow{<\cdot>_\mu}$ Ni potent lie projectors (structure)

Ni potent injectors $\xrightarrow{E(\cdot)}$ Ni potent Lie injectors
$\wr$ $1:1$ $\wr$
Cartan subgroups $\xrightarrow{<\cdot>_\mu}$ Cartan subalgebras

$\underbrace{\qquad\qquad}$ Ascrate $\underbrace{\qquad\qquad}$
groups in   associated
$E(A)$ Lie algebra
            A^0

9.7 Open-ended questions and exercises

Open-ended questions 8 *(i) Are Carter subgroups exactly the nilpotent injectors (also for infinite fields) within unit groups of unitary finite-dimension associative solvable algebras possessing a separable factor algebra by the nilradical?*

(ii) Do Fischer subalgebras exist in arbitrary (solvable) Lie algebras?

(iii) Are the Fischer subalgebras conjugated in arbitrary (solvable) Lie algebras?

(iv) What are the answers of the previous questions for nilpotent Lie injectors and Lie projectors?

(v) Does a general connection between Fischer subalgebras and Fischer subgroups exist for arbitrary associative algebras as described in remark 13?

Excercise 223 *Let G be a semidirect product of a nilpotent normal subgroup and an abelian subgroup. Is every Fischer subgroup of G identical to the Fitting subgroup of G?*

Excercise 224 *True or false: A Fischer-subgroup is maximal nilpotent.*

Excercise 225 *Prove the identity of Dedekind for groups.*

Excercise 226 *Prove the identity of Dedekind for associative algebras.*

Excercise 227 *Prove the identity of Dedekind for Lie algebras.*

Excercise 228 *Let K be a field. Determine the nilpotent injectors of $E(\delta_{u,3})$ and the Lie injectors of $(\delta_{u,3})^\circ$.*

Excercise 229 *Let K be a field. Determine the nilpotent projectors of $E(\delta_{u,3})$ and the Lie projectors of $(\delta_{u,3})^\circ$.*

Excercise 230 *Let K be a field. Determine the Fischer subgroups of $E(\delta_{u,3})$ and the Fischer subalgebras of $(\delta_{u,3})^\circ$.*

Excercise 231 *Let K be a field. Determine the Carter subgroups of $E(\delta_{u,3})$ and the Cartan subalgebras of $(\delta_{u,3})^\circ$.*

Excercise 232 *Let K be a field. Determine the Fitting subgroup of $E(\delta_{u,3})$ and the nilradical of $(\delta_{u,3})^\circ$.*

Excercise 233 *Let K be a field. Determine the nilpotent injectors of $E(K\Pi_3)$ and the Lie injectors of $(K\Pi_3)^\circ$.*

Excercise 234 Let K be a field. Determine the nilpotent projectors of $E(K\Pi_3)$ and the Lie projectors of $(K\Pi_3)^\circ$.

Excercise 235 Let K be a field. Determine the Fischer subgroups of $E(K\Pi_3)$ and the Fischer subalgebras of $(K\Pi_3)^\circ$.

Excercise 236 Let K be a field. Determine the Carter subgroups of $E(K\Pi_3)$ and the Cartan subalgebras of $(K\Pi_3)^\circ$.

Excercise 237 Let K be a field. Determine the Fitting subgroup of $E(K\Pi_3)$ and the nilradical of $(K\Pi_3)^\circ$.

Excercise 238 Let K be a field of characteristic zero. Determine the nilpotent projectors of $E(D_3)$ and the Lie projectors of $(D_3)^\circ$.

Excercise 239 Let K be a field of characteristic zero. Determine the nilpotent injectors of $E(D_3)$ and the Lie injectors of $(D_3)^\circ$.

Excercise 240 Let K be a field of characteristic zero. Determine the Fischer subgroups of $E(D_3)$ and the Fischer subalgebras of $(D_3)^\circ$.

Excercise 241 Let K be a field of characteristic zero. Determine the Carter subgroups of $E(D_3)$ and the Cartan subalgebras of $(D_3)^\circ$.

Excercise 242 Let K be a field of characteristic zero. Determine the Fitting subgroup of $E(D_3)$ and the nilradical of $(D_3)^\circ$.

Chapter 10
Outlook on series III

We have announced that we will analyze within series III in particular the results of chapters 6 to 8 for the Solomon algebras in characteristic zero, the Solomon-Tits algebras and the algebras of upper and lower triangular matrices within the context of associative algebras possessing a self-centralizing radical complement. This presentation includes a general theory for this kind of algebras which is demonstrated at and adjusted on these three algebras. One of the main aims is to classify and determine all maximal Lie nilpotent subalgebras. Besides this we will demonstrate some of the other topics of this work on those algebras.

Solvable group algebras based on non-abelian groups do not possess a self-centralizing radical complement. Therefor an individual chapter is included in series III focussing on this topic. One important question is to analyze the existence of maximal Lie nilpotent subalgebras arising from group algebras based on subgroups. The results are discussed e.g. for dihedral and quaternion groups.

In addition, we focus at the beginning of series III on the compatibility of the results with substructures, tensor products, matrix algebras, factor algebras, inverse algebras, algebras of the form eAe for an idempotent e, zero-extension algebras and algebras with the adjunction of an unit generalizing the results to non-unital algebras.

> 'It is true that a mathematician who is not also something of a poet will never be a perfect mathematician.'

(Karl Weierstraß)

> Outlook on
> Series III

Sun

D_n

VII n

Self-centralizing
radical complement ← How to be verified?

$m_i^c(n)$
$\uparrow^\circ$

$\gamma(A)^\circ$ M avoid?
double   surge of
centralizing isomorphism
               classes

$T = C_A(T)$

chapters 6.9 ?

open what   attackers?
questions?   (schemes?) $Z(n)$? double-centralizing
                                    subgroups/subalgebras

                   ★ non-self-centralizing
                     radical complement
            V 6 software
$A < G$?
   $\text{or } 6$? D_{2n}
   etc? 2 co-minimum? $A 4$? A^- ?

effect?
non-real
d-nilpotery
class?

non-real
dimension?

$A \otimes B$?

List of Tables

3.1 nilpotency and solvable classes linked to $K\Pi_n$ and $(K\Pi_n)^\circ$. 47
3.2 nilpotency and solvable classes linked to D_n and $(D_n)^\circ$ 48
3.3 nilpotency and solvable classes linked to $E(K\Pi_n)$ 49
3.4 nilpotency and solvable classes for $E(D_n)$ 50
3.5 nilpotency and solvable classes of $\delta_{u,n}$ and $\delta_{o,n}$ 51

List of Figures

the theorem of Xiankun Du 30
standard examples 31

connections for solvability 53

Carter subgroups and Cartan subalgebras 73
Carter subgroups of standard examples 74

the Fitting subgroup and the Lie nilradical 82
the Fitting subgroup of standard examples 83

associative closure of maximal Lie nilpotent subalgebras 89
maximal Lie nilpotency: manifold centralizers 102
maximal Lie nilpotency: attractor and repeller properties 103
futile algebras 110
unitalization map 114
maximal Lie nilpotency: finite number of isomorphism classes . . . 117
maximal Lie nilpotency: cardinalities 120

maximal Lie nilpotency: the correspondence theorem 135

maximal nilpotent subgroups: manifold centralizers 147
maximal nilpotent subgroups: attractor and repeller properties . . 148
maximal nilpotency of the group of units: finite number of isomorphic classes 153

extreme maximal nilpotent substructures 170

outlook on series III 175

Bibliography

[1] M. D. Atkinson, Solomon's descent algebra revisted, Bull. London Math. Soc. 24, 1992, 545-551

[2] S. A. Amitsur, Finite subgroups of division rings, Trans. Amer. Math. Soc. 80, 1955, 361-386

[3] D. W. Barnes, On Cartan subalgebras of Lie algebras, Math. Z. 101, 1967, 350-355

[4] T. Bauer, Über die Struktur der Solomon-Algebren, Bayreuther Mathematische Schriften, Heft 63, 2001, 1-102

[5] T.Bauer, S. Siciliano, Carter subgroups in the group of units of an associative algebra, Bulletin of the Australian Mathematical Society / Volume 71 / Issue 03 / June 2005, 471-478

[6] N. Bourbaki, Lie Groups and Lie Algebras, Chapters 7-9 (Elements of Mathematics), Springer-Verlag, 2008

[7] G. Benkart, Cartan subalgebras in Lie algebras of Cartan type, Canadian Mathematical Society Conference Proceedings 5, 1986, 157-187

[8] Yakov Berkovitch - Avinoam Mann, On sums of degrees of irreducible characters, Journal of Algebra, 199, 1998, 646-665

[9] N. Bourbaki, Groupes et algèbres de Lie, Chapitre VII, Hermann, Paris, 1973

[10] A.A. Bovdi - I.I. Khripta, Generalized Lie nilpotent group rings, Math. USSR Sbornik 57(1), 1987, 165-169

[11] Carter, R., On a class of finite soluble groups, Proc. London Math. Soc. 9, 1959, 623-640

[12] R. Dedekind, Über Gruppen, deren sämmtliche Theiler Normaltheiler sind, Mathematische Annalen, Band 48,4, 1897, 548-561

[13] Di Martino, L. - Tamburini, M.C. - Zalesskiĭ, A.E., Carter subgroups in classical groups, J. London Math. Soc. 55, 1997, 264-276

[14] Xiankun Du, The centers of a radical ring, Canad. Math. Bull. 35, no. 2, 1992, 174-179

[15] Gutan, Marin - Kisielewicz, Andrezej, Reversible group rings, Journal of algebra 279, 2004, 280-291

[16] Gorenstein, D., Finite Groups, Harper and Row, New York, 1968

[17] Hallahan, C.H. - Overbeck, J, Cartan subalgebras of meta-nilpotent Lie algebras, Math. Z. 116, 1970, 215-217

[18] G. H. Hardy - E. M. Wright, An Introduction To The Theory Of Numbers, Oxford University Press, U.S.A., 6th edition., 2009

[19] I.N. Herstein, Finite multiplicative subgroups in division rings, Pacific J. Math. 3, 1953, 121-126

[20] I.N. Herstein, Topics in Ring Theory, University of Chicago Press, Chicago, 1969

[21] I.N. Herstein, Lie and Jordan structures in simple associative rings, Bull. AMS, Vol. 67, No. 6, 1961, 517-531

[22] I.N. Herstein, On the Lie structure of an associative ring, Journal of algebra, Vol. 14, Issue 4, April 1970, 561-571

[23] L.K. Hua, Some properties of s-fields, Proc. Nat. Acad. Sci. U.S.A. 35, 1949, 533-537

[24] B. Huppert, Endliche Gruppen I, Springer-Verlag, Berlin, 1967

[25] I.M. Isaacs, Algebra, a graduate course, Brooks/Cole Publishing Company, Pacific Grove, California, 1993.

[26] G. Ivanyos, Finding the radical of matrix algebras using Fitting decomposition, J. Pure Appl. Algebra 139, 1999, 159-182

[27] N. Jacobson, Lie Algebras, Wiley Interscience, New York London, 1962

[28] N. Jacobson, Schur's theorems on commutative matrices, Bull. Amer. Math. Soc. Volume 50, Number 6, 1944, 431-436

[29] S.A. Jennings, Central chains of ideals in an associative ring, Duke Math. Journal 9, 1942 341-355

[30] S.A. Jennings, Radical rings with nilpotent associated group. Trans. Roy. Soc. Can. ser. III 49, 1955, 31-38

[31] Armin Jöllenbeck, Abgeleitete Algebren, Mathematisches Seminar der Christian-Albrechts-Universität zu Kiel, Diplomarbeit, 1994

[32] G. Karpilovsky, The Jacobson Radical Of Group Algebras, Elsevier, Amsterdam, 1987

[33] Gregory Karpilovsky, Unit groups of classical rings, Clarendon Press, Oxford, 1988

[34] Adalbert Kerber, Zu einer Arbeit von J. L. Berggren über ambivalente Gruppen, Pacific Journal of Mathematics, Vol. 33, No. 3, 1970

[35] I.I. Khripta, The nilpotence of the multiplicative group ring, Mat. Zametki 11, 1972, 191-200

[36] Max-Albert Knus et.a., The book of involutions, AMS Colloquium Publications, Volume 44, 1998

[37] Michiel Kosters, Algebras with only finite many subalgebras, http://arxiv.org/pdf/1401.1607.pdf

[38] Robert L. Kruse - David T. Price, Nilpotent rings, Gordon and Breach Science Publishers Ltd., 1969

[39] T. Y. Lam, Finite Groups Embeddable in Division Rings, www.arxiv.org

[40] H. Laue, Assoziative Algebren, Vorlesung am Mathematischen Seminar der CAU zu Kiel, WS 2010/2011

[41] H. Laue, On the associated Lie ring and the adjoint group of a radical ring, Canadian mathematical bulletin, Vol. 27, No. 4, 1984, 217 ff.

[42] H. Laue, Lie-Algebren, Vorlesung am Mathematischen Seminar der CAU zu Kiel, WS 2008/2009

[43] Macdonald, I. G., Symmetric functions and Hall polynomials. Second edition. Claren- don Press, Oxford, 1995

[44] K. Motose, On the nilpotency index of the radical of a group algebra III, J. London Math. Soc 2, 25, 1982, 39-42

[45] K. Motose, On the nilpotency index of the radical of a group algebra IV, Math. J. Okayama Univ. 25, 1983, 35-42

[46] K. Motose, On the nilpotency index of the radical of a group algebra V, Journal of Algebra 90, 1984, 251-258

[47] I. B. S. Passi - D. S. Passman - S. K. Sehgal, Lie solvable group rings, Can. J. Math., Vol. XXV, No. 4, 1973, 748-757

[48] D. S. Passman, Observations on group rings, Communications in Algebra, 5(11), 1977, 1119-1162

[49] S. Perlis - G.L. Walker, Abelian group algebras of finite order, Trans. Amer. Math. Soc. 68, 1950, 420-426

[50] R.S. Pierce, Associative Algebras, Springer-Verlag, New York, 1982

[51] Robinson, D.J.S., A course in the theory of groups, Springer-Verlag, New York, 1982

[52] Robinson, Geoffrey R., A bound on norms of generalized characters with applications, Journal of Algebra, 212, 1999, 660-668

[53] Robinson, Geoffrey R., More bounds on norms of generalized characters with applications to p-local bounds and blocks, Bulletin of London Mathematical Society, 37, 4, 2005, 555-565

[54] Robinson, Geoffrey R., On generalized characters of nilpotent groups, Journal of Algebra, 308, 2007, 822-827

[55] Robinson, Geoffrey R., On the minimal norm of a non-regular generalized character of an arbitrary finite group, Bulletin LMS, 2010

[56] Scheja, Günter - Storch, Uwe, Lehrbuch der Algebra, Teil 2, Teubner-Verlag, Stuttgart, 1988

[57] Manfred Schocker, The module structure of the Solomon-Tits algebra of the symmetric group, J. Alg. 301(2006), No.2, pages 554-586 (Peprint available at http://arxiv.org/abs/math/0505137)

[58] Scholz, Karsten, Zentralreihen in Radikalringen, Diplomarbeit, Mathematisches Seminar der CAU zu Kiel, 1996

[59] Jean-Pierre Serre, Linear Representations of Finite Groups (Graduate Texts in Mathematics), Springer, 1996

[60] A. Shalev, The derived length of Lie soluble group rings I, J. Pure Appl. Algebra, 78, 1992, 291-300

[61] A. Shalev, The derived length of Lie soluble group rings II, J. London Math. Soc., 49, 1994, 93-99

[62] S. Siciliano, Cartan subalgebras in Lie algebras of associative algebras, Communications in Algebra, Volume 34, Issue 12 December 2006, 4513-4522

[63] S. Siciliano, On the Cartan subalgebras of Lie algebras over small fields, J. Lie Theory 13, 2003, 511-518

[64] Benjamin Steinberg, http://math.stackexchange.com/questions/819466/

[65] Stitzinger, E., Theorems on Cartan subalgebras like some on Carter subgroups, Trans. Amer. Math. Soc. 159, 1971, 307-315

[66] Stitzinger, E., Minimal non-nilpotent solvable Lie algebras, Proceedings of the AMS, Volume 28, No. 1, April 1971

[67] H. Strade - R. Farnsteiner, Modular Lie algebras and their representations. Marcel Dekker, New York, 1988

[68] Tamburini, M.C. - Vdovin, E.P., Carter subgroups in finite groups, J. Algebra 255, 2002, 148-163

[69] M. Theede, Die aufsteigende Zentralreihe in Einheitengruppen modularer Gruppenalgebren für Klassen metabelscher p-Gruppen, Dissertation, Kiel, 2016

[70] Hisao Tominaga, Algebras with only finitely many subalgebras, Mathematical Journal of Okayama University, Volume 19, Issue 1 1976 Article 8, December 1976

[71] Vinroot, C. Ryan, Twisted Frobenius-Schur indicators of finite symplectic groups, Journal of Algebra, 293, 2005, 279-311

[72] Vinroot, C. Ryan, A note on orthogonal similitude groups, Linear and Multilinear Algebra, 54(6), 2006, 391-396

[73] Vinroot, C. Ryan, Character degree sums and real represenations of finite classical groups of odd characteristic, Journal of Algebra and Its Applications, 09, 633 (2010), 633-658

[74] Dalla Volta, F. - Lucchini, A. - Tamburini, M.C., On the Conjugacy Problem for Carter Subgroups, Comm. Algebra 26, 1998, 395-401

[75] S. Wirsing, Separabilität in kommutativen und auflösbaren Algebren, Unter Berücksichtigung nicht-unitärer assoziativer Algebren, Disserta-Verlag, 2015, Hamburg

[76] S. Wirsing, Über Einheitengruppen modularer Gruppenalgebren, Disserta-Verlag, 2015, Hamburg

[77] S. Wirsing, Über die Struktur der Solomon-Tits-Algebren der symmetrischen Gruppe, Eine Analyse assoziativer, gruppentheoretischer und Lie-theoretischer Phänomene, Disserta-Verlag, 2015, Hamburg

[78] S. Wirsing, Maximal nilpotente Teilstrukturen I, Nilradikale und Cartan-Teilalgebren in assoziierten Lie-Algebren, Disserta-Verlag, 2015, Hamburg

[79] S. Wirsing, Maximal nilpotent substructure I, Nilradicals and Cartan subalgebras in associative algebras, Disserta-Verlag, 2016, Hamburg

[80] Wikipedia, information for several mathematicians and for the primitive element theorem

[81] R.L. Wilson, Cartan subalgebras of simple Lie algebras, Trans. Amer. Math. Soc 234, 1977, 435-446

[82] D.J. Winter, On the toral structure of p-algebras, Acta Math. 123 (1969) 70-81

[83] The GAP Group, GAP — Groups, Algorithms, and Programming, Version 4.2; Aachen, St Andrews, 1999. (http://www-gap.dcs.st-and.ac.uk/ gap)

[84] http://page.math.tu-berlin.de/ kant/Algebra/algalg3.pdf

[85] http://de.wikipedia.org/wiki/BellscheZahl

[86] http://mathoverflow.net/questions/206602/are-carter-subgroups-nilpotent-projectors/206608

[87] https://math.stackexchange.com/questions/276627/a-finite-field-extension-that-is-not-simple

Index

$(A \times A; \odot)$ zero extension of A, 17
$(A^\circ)^{(n)}$ lower Lie central chain, 22
$(G^{[n]})_{n \in \mathbb{N}}$ descending chain of commutators, 37
$(K; L)$ field extension, 18
$(L^{[n]})_{n \in \mathbb{N}}$ descending chain of Lie commutators, 37
$A(a, b)$ generalized quaternion algebra, 18
A/I factor algebra, 16
$A/rad(A)$ factor algebra by the nilradical, 18
A^K adjunction of an unit, 17
A° associated Lie algebra, 17
$A^{n \times n}$ algebra of $n \times n$-matrices over A, 17
A^{op} or A^- opposite or inverse algebra of A, 17
A_n alternating group of degree n, 15
$Aug(KG)$ augmentation ideal of KG, 17
$B(n)$ Bell numbers, 28
C_n or Z_n cyclic group of order n, 15
$D(\alpha)$ defect class, 27
$D(n, K)$ - the set of diagonal matrices in $K^{n \times n}$, 17
D_n Solomon algebra, 19
D_{2n} dihedral group of order $2n$, 15
$E(A)$ group of units of A, 15
$GF(p^n)$ finite field possessing p^n elements, 18
$GL(n, q)$ general linear group of degree n over $GF(q)$, 15
$J(A)$ Jacobson radical of A, 19
$K(a)$ smallest subfield in L containing a and K, 18

$K(t)$ field of fractions over K in the variable t, 18
$K(t_1, \ldots, t_n)$ field of fractions over K the variables $t_1, \ldots, t_n$, 18
KG group algebra, 16
KM monoid algebra, 16
$K[a]$ smallest subalgebra containing a and K, 17
$K[t]$ polynomial algebra over K in the variable t, 17
$K[t_1, \ldots, t_n]$ polynomial algebra over K the variables $t_1, \ldots, t_n$, 17
$K\Pi_n$ Solomon-Tits algebra, 19
$Q(A)$ quasi regular group, 22
$Q(A)$ quasiregular group of A, 16
$Q(A)^{(n)}$ lower central chain of $Q(A)$, 23
Q_{4n} quaternion group of order $4n$, 15
$S(n, k)$ Stirling numbers, 28
SD_{2^n} semi-dihedral group of order 2^n, 15
S_n symmetric group of degree n, 15
$W_n(Q(A))$ special central chain, 23
$X_n(Q(A))$ central chain related to additive group, 23
$Y_n(Q(A))$ upper central chain of $Q(A)$, 23
$Z_n(A^\circ)$ upper Lie central chain, 22
$[\cdot, \cdot]$ commutator, 23
$\circ$ Lie composition, 22
$\delta_{o,n}$ algebra of upper triangular matrices of $K^{n \times n}$, 19
$\delta_{u,n}$ algebra of lower triangular matrices of $K^{n \times n}$, 19
$\gamma_k(G)$ k-th member of the lower central chain, 23

$\langle T \rangle_K$ K-linear span, 17
$\langle T \rangle_{A_1}$ unital subalgebra generated by T, 17
$\langle T \rangle_A$ subalgebra generated by T, 17
$\ltimes$ semidirect product of algebras, 16
$\ltimes$ semidirect product of groups, 16
$\mathbb{C}$ complex number field, 18
$\mathbb{H}$ real quaternion algebra, 18
$\mathbb{N}$ natural numbers, 15
$\mathbb{N}_0$ natural numbers containing zero, 15
$\mathbb{Q}$ rational number field, 18
$\mathbb{R}$ real number field, 18
$\mathbb{Z}$ the set of integers, 17
ω_d primitive dth root of unity, 18
$\oplus$ direct sum of algebras, 16
$\otimes$ tensor product of algebras, 16
$\star$ star composition, 22
$\times$ direct products of algebras, 16
$\times$ direct products of groups, 16
φ_T special function, 25
$\wedge_n, \wedge$ product on Π_n, 28
$a^{(b)}$ star conjugate element, 23
a^- star inverse, 23
$cl(Q(A))$ class of nilpotency of $Q(A)$, 23
$cl(\cdot)$ class of nilpotency, 22
eAe identical to $\{eae \mid a \in A\}$ for an idempotent e, 17
$gl(n,K)$ identical to $(K^{n\times n})^\circ$, 17
n_K - identical to $\sum_{i=1}^{n} 1_K$, 17
$p(n)$ number of partitions of n, 27
$rad(A)$ nilradical of A, 19
$s\delta_{o,n}$ algebra of strict upper triangular matrices of $K^{n\times n}$, 19
$s\delta_{u,n}$ algebra of strict lower triangular matrices of $K^{n\times n}$, 19
$st(A)$ solvable class of A, 38
$st(G)$ solvable class of G, 37
$st(L)$ solvable class of L, 37
$\binom{T}{i}$ T choose i, 25

Abel, Niels, 103

algebra
  associative powers, 16
  centralizer, 16
  class of associative algebras, 16
  class of associative unitary algebras, 16
  class of nilpotency, 16
  derivation of A, 16
  dimension, 16
  generalized Jordan decomposition, 16
  solvable class, 16
  sum of elements, 16
Artin, Emil, 42
associated Lie algebra
  class of nilpotency, 22
  definition, 22
  lower central chain, 22
  nilpotent, 22
  upper central chain, 22
associative algebra
  class of solvability, 38
  definition of solvability, 37
  descending chain of commutators, 38
  Jacobson radical, 21
  Lie solvable, 38, 40
  nil, 22
  nilradical, 21
  radical algebra, 22
  solvability of related structures, 42
  solvability of the group of units, 42
  tensor products and solvability, 41
Atkinson, Michael D., 27

base matrices of $K^{n\times n}$, 18
Bauer, Thorsten, 7, 9, 27, 59–65, 68–70
Bell, Eric Temple, 28, 105
Borel, Armand, 9, 10, 19, 43, 44
Bovdi, Adalbert, 8

Cartan subalgebra
    Carter subgroup, 64
    solvable associative algebra, 59
Cartan, Elie, 7, 43, 59, 64, 67–71, 74–76, 85, 87, 97, 98, 100, 114, 115, 120, 122, 123, 129, 132, 155, 165, 166
Carter, Roger, 7, 9, 11, 59–65, 67–71, 74–76, 132, 143–145, 148, 153, 156, 159, 165, 166, 170, 171
characteristic of K, 17
circle group
    Du, 24
    Jennings, 23
    Laue, 24
    main theorem, 24
    operation on the additive group, 23
correspondence for maximal abelian substructures, 135
correspondence of attraction sections, 157

Dedekind, Richard, 167
determinant, 18
Du, Xiankun, 8, 11, 21, 23, 24, 26–28, 31, 33, 37, 70, 71, 78, 80, 132, 133, 145, 156

Engel, Friedrich, 87, 127, 158

Fischer subalgebra
    definition, 164
    nilradical, 164
Fischer subgroups
    definition, 163
    Fitting subgroup, 163
Fischer, Bernd, 11, 163, 164
Fitting, Hans, 14, 19, 60, 62, 77–80, 83, 132, 143–145, 148, 153, 155, 163, 164, 167, 168
futile algebras
    adjunction of an unit, 105
    Bell numbers, 106
    characterizations, 104
    definition, 103
    Galois theory, 103
    Goursat-lemma, 104
    ideals and idempotents of K^n, 106
    separable commutative algebra, 107
    unital and non-unital, 105
    unital subalgebras of K^n, 106
    upper bound for subalgebras, 108

Galois, Evariste, 70, 103
Goursat, Jean-Baptiste, 104, 127
group
    ascending central chain, 14
    centralizer, 14
    circle composition, 14
    class of groups, 14
    class of solvability, 37
    commutator, 14
    commutator series, 14
    conjugation, 14
    derived series, 14
    derived subgroup, 14
    descending central chain, 14
    descending sequence of derivations, 37
    exponent, 15
    factor group, 15
    Fitting length, 77
    Fitting series, 77
    Fitting subgroup, 77
    fully-separable factor, 15
    inverse element, 14
    maximal nilpotent subgroups, 15
    nilpotency class, 14
    normalized units, 14
    normalizer, 14
    order of $\mathcal{G}(U)$, 15
    p-core, 15
    quasiregular conjugate, 15
    quasiregular inverse, 15
    solvable class, 14
    solvable subgroups, 15
    star composition, 14

subgroups possessing the double centralizer-property, 15
subgroups possessing the double-centralizer property, 15
unipotent factor, 15
group algebra
   Carter subgroups and Cartan subalgebras, 67
   Fitting subgroup, 80
   solvability, 45
group of units of a solvable associative algebra
   number of p'-Hall subgroups, 67
   number of Carter subgroups, 67
group of units of an associative solvable algebra
   p-Sylow subgroup, 65
   p'-Hall subgroups, 65
   Carter subgroups, 65

Hall, Philipp, 65–68, 70, 71, 74–76, 80

Jacobson, Nathan, 21, 23, 129
Jennings, Stephen Arthur, 8, 23, 24
Jordan, Marie Ennemond Camille, 16, 83, 86, 87, 129, 137, 158, 160

Köthe, Gottfried, 21
Kosters, Michiel, 104

Laue, Hartmut, 8, 9, 21, 23, 24, 44
Lie algebra
   adjoint representation, 19
   ascending central chain, 20
   associated Lie algebra, 19
   associated Lie composition, 19
   Borel subalgebra, 19
   class of Lie algebras, 20
   class of solvability, 37
   commutator series, 20
   derivation, 20
   derived series, 20
   descending central chain, 20
   descending chain of group commutators, 37
   Fitting null component, 19
   maximal nilpotent Lie subalgebras, 20
   nilpotency class, 20
   nilradical, 20
   order of $\mathcal{M}(T)$, 20
   sequence of substructures, 20
   solvable class, 19
   subalgebras possessing the double-centralizer property, 20
   unit group and K-space creation, 20
Lie, Sophus, 7, 19, 43, 44, 53, 59, 85, 86, 89, 114, 117, 131, 132, 137, 163

Malcev, Anatoli Iwanowitsch, 56, 63, 65, 89, 99, 117, 144, 150
maximal Lie nilpotency
   associative closure, 87
   associative subalgebra, 87
   attractor of Cartan subalgebras, 98
   attractor of the nilradical, 98
   attractor properties, 95
   cardinalities, 117
   centralizer property, 89
   change of radical complement, 99
   characterization by centralizers, 93
   correspondence to the group of units, 132
   double centralizers, 90, 93
   double-centralizer property, 89
   examples, 120
   generation of maximal members, 92
   generation of maximal members II, 94
   main theorem, 122
   maximal subalgebras, 120
   minimal non-nilpotent, 120

nilpotency class, 100
relationships of cardinalities, 117
repeller properties, 95
sequences of double centralizers, 97
maximal Lie-nilpotency
  finite many classes of isomorphism, 114
maximal nilpotency of the group of units
  a direct decomposition, 137
  attracting section of Carter subgroups, 144
  attracting section of Fitting subgroup, 144
  attractors, 141
  cardinalities, 150
  centralizer properties, 138
  change of the radical complement, 144
  characterization by centralizers, 140
  class of nilpotency, 145
  connections of cardinalities, 150
  correspondence to the Lie algebra, 132
  creation of maximal nilpotent subgroups, 139
  creation of maximal nilpotent subgroups II, 140
  creation of maximal nilpotent subgroups III, 140
  double-centralizer properties, 138
  finiteness of the number of isomorphism classes, 148
  fully separable factor, 137
  main theorem, 155
  maximal subgroup, 153
  minimal non-nilpotent, 153
  repeller, 141
  sequences of double centralizers, 142
  triangular matrices, 153
  unipotent factor, 137
  upper bound, 148
Motose, Kaoru, 80

n-tuple space over K, 17
nilpotent injector
  definition, 167
  Fitting subgroup, 167
nilpotent Lie injector
  definition, 168
  nilradical, 168
nilpotent Lie projector
  Cartan subalgebras, 166
  definition, 166
nilpotent projector
  Carter subgroups, 165
  definition, 165
Noether, Emmy, 70
numbers
  Bell number, 14
  binomial coefficient, 14
  equivalent, 14
  factorial of n, 13
  modulo, 14
  partion function, 13
  Stirling number, 14

outlook on series II, 173

primitive orthogonal idempotents in $D(n,K)$, 17

quasi regular group
  commutator, 23
  conjugation, 23
  definition, 22
  inverse, 23
  lower central chain, 23
  nilpotency class, 23
  upper central chain, 23

Schocker, Manfred, 28
Scholz, Karsten, 21, 24
Schur, Issai, 68
Scott, Raymond William, 42, 70
set of fully-separable elements, 17

sets
    cartesian product, 13
    difference, 13
    hook, 13
    hook with zero, 13
    intersection, 13
    order, 13
    power set, 13
    the empty set, 13
    the set of subsets of order i of T, 14
    union, 13
Siciliano, Salvatore, 9, 59–65, 68–70
Skolem, Albert Thoralf, 70
Solomon algebra
    Cartan subalgebras, 71
    Carter subgroups, 71
    class of nilpotency, 27
    defect class, 16
    Fitting subgroup, 79
    solvable class, 46
Solomon, Louis, 12, 27, 28, 46, 71, 74, 79, 80, 83, 106, 120, 173
Solomon-Tits algebra
    Cartan subalgebras, 71
    Carter subgroups, 71
    class of nilpotency, 28
    composition on ordered set partitions, 15
    Fitting subgroup, 80
    ordered set partitions, 15
    solvable class, 46
solvable algebra
    Carter subgroup, 62
    conjugacy of Carter subgroups, 63
    Fitting subgroup, 78
star group
    ascending central chain
        elementary-abelian factors, 26
Stirling, James, 28
Stitzinger, E., 120
sums of idempotents, 17

Sylow, Peter Ludwig Mejdell, 45, 65–71, 74, 80

Tits, Jacques, 12, 28, 46, 71, 74, 80, 83, 106, 120, 173
Tominaga, Hisao, 103
trace, 18
Triangular matrices
    Fitting subgroup, 79
triangular matrices
    Carter and Cartan, 70
    nilpotency class, 26
    solvable class, 46

unordered set partitions, 28

van der Waerden, Bartel Leendert, 22

Wedderburn, Joseph, 42, 56, 63, 65, 89, 99, 117, 144, 150
Weierstraß, Karl, 173

Zassenhaus, Hans Julius, 68
zero extension, 56